Peter Amsler

Digital In-line Holographic Microscope for Ice Crystals

Peter Amsler

Digital In-line Holographic Microscope for Ice Crystals

Ice crystal habits and their optical properties

Südwestdeutscher Verlag für Hochschulschriften

Impressum/Imprint (nur für Deutschland/ only for Germany)
Bibliografische Information der Deutschen Nationalbibliothek: Die Deutsche Nationalbibliothek verzeichnet diese Publikation in der Deutschen Nationalbibliografie; detaillierte bibliografische Daten sind im Internet über http://dnb.d-nb.de abrufbar.

Alle in diesem Buch genannten Marken und Produktnamen unterliegen warenzeichen-, marken- oder patentrechtlichem Schutz bzw. sind Warenzeichen oder eingetragene Warenzeichen der jeweiligen Inhaber. Die Wiedergabe von Marken, Produktnamen, Gebrauchsnamen, Handelsnamen, Warenbezeichnungen u.s.w. in diesem Werk berechtigt auch ohne besondere Kennzeichnung nicht zu der Annahme, dass solche Namen im Sinne der Warenzeichen- und Markenschutzgesetzgebung als frei zu betrachten wären und daher von jedermann benutzt werden dürften.

Verlag: Südwestdeutscher Verlag für Hochschulschriften Aktiengesellschaft & Co. KG
Dudweiler Landstr. 99, 66123 Saarbrücken, Deutschland
Telefon +49 681 37 20 271-1, Telefax +49 681 37 20 271-0
Email: info@svh-verlag.de
Zugl.: Zürich, ETH, Diss., 2009

Herstellung in Deutschland:
Schaltungsdienst Lange o.H.G., Berlin
Books on Demand GmbH, Norderstedt
Reha GmbH, Saarbrücken
Amazon Distribution GmbH, Leipzig
ISBN: 978-3-8381-1461-3

Imprint (only for USA, GB)
Bibliographic information published by the Deutsche Nationalbibliothek: The Deutsche Nationalbibliothek lists this publication in the Deutsche Nationalbibliografie; detailed bibliographic data are available in the Internet at http://dnb.d-nb.de.

Any brand names and product names mentioned in this book are subject to trademark, brand or patent protection and are trademarks or registered trademarks of their respective holders. The use of brand names, product names, common names, trade names, product descriptions etc. even without a particular marking in this works is in no way to be construed to mean that such names may be regarded as unrestricted in respect of trademark and brand protection legislation and could thus be used by anyone.

Publisher: Südwestdeutscher Verlag für Hochschulschriften Aktiengesellschaft & Co. KG
Dudweiler Landstr. 99, 66123 Saarbrücken, Germany
Phone +49 681 37 20 271-1, Fax +49 681 37 20 271-0
Email: info@svh-verlag.de

Printed in the U.S.A.
Printed in the U.K. by (see last page)
ISBN: 978-3-8381-1461-3

Copyright © 2010 by the author and Südwestdeutscher Verlag für Hochschulschriften Aktiengesellschaft & Co. KG and licensors
All rights reserved. Saarbrücken 2010

Contents

1 Introduction   1
    1.1 Ice crystal growth and habit formation 3
    1.2 Outline of this thesis 7

2 Basic optical concepts of analog and digital holography   9
    2.1 Coherence length 9
    2.2 Depth of field and circle of confusion 11
    2.3 Aberrations and distortions 12
       2.3.1 Astigmatic aberration 13
       2.3.2 Coma aberration 14
       2.3.3 Spherical aberration 15
       2.3.4 Distortions 16

3 Theory of operation of analog and digital holography   19
    3.1 The Kirchhoff-Helmholtz (KH) transformation 19
    3.2 Intensity distribution of the diffraction pattern 25
    3.3 Resolution of the camera 27
    3.4 Energy density on the camera chip 28
    3.5 Light absorption of water and ice 29

4 The holographic microscope HOLIMO   31
    4.1 Abblation due to deposited energy 32

4.2	Camera laser trigger	33
4.3	Ice crystal fall speed	34
4.4	Application of the KH transformation	35
4.5	Resolution considerations of the microscope	39
	4.5.1 Edge blurring	42
4.6	Particle hitting rate	43
4.7	Data processing	44
4.8	HOLIMO I setup	52
4.9	HOLIMO II setup	53

5 Proof of concept studies — 57

5.1	USAF target for HOLIMO I	57
5.2	PSL targets for HOLIMO I	59
5.3	Ice analogue targets for HOLIMO I	60
5.4	Tests with HOLIMO II	63
5.5	ZINC measurements with HOLIMO II	64

6 Experimental and modelling results from AIDA measurements — 67

6.1	AIDA facility	67
	6.1.1 WELAS	68
	6.1.2 Depolarization Instrument	70
6.2	AIDA Experiment IN11_2 from November 2007	72
6.3	More experiments from the IN11 campaign from November 2007	79
	6.3.1 Heterogeneous cloud experiments from IN11	79
	6.3.2 Homogeneous cloud experiments from IN11	83

7 Experimental results from PINC II field measurements — 89

7.1	Introduction	89
7.2	Experiment and setup	91
7.3	Preliminary results	95

8	3 D modelling of ice crystal habits	99
9	Conclusion	103
	Acknowledgements	105
	List of publications	107
	Tables and figures	108
	Bibliography	122
A	List of Acronyms	141
B	List of Symbols	143
C	Image acquisition	145
D	Commercial software	147
E	Basic reconstruction routine in Matlab	149
	E.1 Negative vs. positive reconstructions	151
	E.2 Contrast image	152

Glück ist, wenn Zufall auf Vorbereitung trifft
anonymous

Abstract

Earth's radiative budget is very sensitive to changes in cloud cover in the upper troposphere. There are indications that global cirrus coverage is increasing with time, probably due to increased air traffic because aerosols and gases are directly emitted into the most sensitive region. Cirrus clouds are the dominant cloud type in the upper troposphere, covering approximately 20 to 30% of Earth's surface. They consist of varying particles of various shapes in a broad size range from tens to several hundred μm that can absorb and scatter solar and terrestrial radiation in different ways depending on their habit and orientation. The particles of cold and mixed phase clouds can emerge from homogeneous and heterogeneous nucleation. For this reason, the formation and growth of cirrus particles depend on both anthropogenic and natural aerosols, temperature and relative humidity.

This thesis discusses the technical part of the digital in-line holographic microscope HOLIMO supported by proof of concept studies performed in the laboratory. Several target objects were used for calibration and improvement. Additionally, images were taken from ice analogues in order to compare with images obtained by different instruments like scanning electron microscope. The measurements of HOLIMO and scanning electron microscope agreed within 5% and less. A multitude of computer programs were developed for the purpose of object reconstruction from holographic interference pattern and subsequent characterization.

Aberrations and distortions of optical systems is another chapter in this thesis, but the main part concentrates on data evaluation of three measurement campaigns and the description of the used instruments within. The first and second measurement campaign took place at the AIDA chamber from the Institute for Meteorology and Climate Science at the Research Center Karlsruhe in Germany in November/December 2007 and December 2008. The objectives were to investigate cold and mixed phase cloud evolutions by means of hydrometeor habit classification in general and the relationship between ice crystal habit and light depolarization especially.

Warm, mixed phase and cold clouds have been produced inside AIDA. Sizes and habits of ice crystals and droplets were related to existing inorganic aerosol type, relative humidity with respect to ice (RH_{ice}), temperature (T) and experiment time in different experiments. In experiments 1 and 3, which were initiated from supercooled water and supercooled water plus sulfuric acid respectively in the first campaign in Karlsruhe, regular crystal shapes dominated. The hydrometeors detected with HOLIMO were within a size range of 2 to 140 μm (average size of 11 μm) and 2 to 50 μm (average size of 9 μm) respectively. In contrast, experiment 2 of the same campaign was initiated with supercooled water and ice seeds in a similar T and RH_{ice} range. As a result, irregular ice crystals within a size range of 2 to 168 μm (average size of 17 μm) became much more dominant. Experiment 4 of this campaign was initiated at much colder temperatures in order to examine homogeneous freezing of sulfuric acid. Only aggregates were found within the resolution of the holographic microscope in a size range from 2 to 43 μm (average size of 12 μm). It was found for experiment 2 especially that very thin plates exhibit very low linear depolarization values down to 0.04. This value is below reported values from LIDAR measurements for cloud droplets.

Mixed phase clouds were investigated at the high altitude research station on the Jungfraujoch. The question was if mixed-phase clouds are well-mixed or if isolated patches of only ice crystals and only cloud droplets exist. The preliminary results of our experiment reveal a different behavior in the frequency of occurrence of numbers of ice crystals or cloud droplets from a modelled well-mixed cloud.

Zusammenfassung

Der Strahlungshaushalt der Erde ist sehr empfindlich gegenüber Veränderungen in der Wolkenbedeckung in der oberen Troposphäre. Es gibt Hinweise, dass die globale Zirrenbedeckung mit der Zeit zunimmt. Die Ursache hierfür könnte die Zunahme vom Luftverkehr sein, denn Aerosole und Gase werden direkt in die empfindlichste Zone emittiert. Zirruswolken sind dominant in der oberen Troposphäre. Sie bedecken ungefär 20 bis 30% der Erdoberfläche. Sie bestehen aus den verschiedensten Partikel mit unterschiedlichen Grössen von einigen zehn bis mehreren hundert μm. Sie können solare und terrestrische Strahlung absorbieren und reflektieren und sie tun dies auf unterschiedlichste Weise, abhängig von ihrem Habitus und ihrer Orientierung. Die Wolkenpartikel von kalten und gemischtphasigen Wolken können aus homogenen und heterogenen Gefrierprozessen entstehen, ihre Entstehung und ihr Wachtum aber werden beinflusst sowohl durch anthropogene als auch durch natürliche Aerosole, die Temperatur und den zur Verfügung stehenden Wasserdampf in der Atmosphäre.

Diese Arbeit diskutiert den technischen Aspekt vom digitalen in-line holographischen Mikroskope HOLIMO unterstützt von proof of concept Studien durchgeführt im Labor. Unterschiedlich Zielobjekte wurden benutzt, um Kalibrierungen und Verbesserungen vorzunehmen. Zusätzlich wurden Bilder von Eisanalogen angefertigt und mit den Resultaten von anderen Bildgebungsverfahren wie Rasterelektronenmikroskop verglichen. Die Übereinstimmung der Messungen mit HOLIMO und mit Rasterelektronenmikroskop weist eine Abweichung von nicht mehr als 5% aus. Eine Vielzahle von Komputerprogrammen wurde entwickelt, um Objektrekonstruktion aus holographische Interfrenzen mit anschliessender Charakterisierung zu betreiben.

Abbildungsfehler und -verzerrungen von optischen Systemen werden in einem weiteren Kapitel dieser Arbeit besprochen. Der Hauptteil aber konzentriert sich auf die Auswertung von Daten von drei unterschiedlichen Kampagnen und die dort verwendeten Instrumente. Die erste und zweite Kampagne fanden statt bei der AIDA Kammer vom Institute für Meteorologie und Klimawissenschaften am Forschungszentrum in Karlsruhe in Deutschlang im November/Dezember 2007 und Dezember 2008. Ziele waren die Untersuchung vom Habitus von Hydrometeoren von kalten und gemischtphasigen Wolkenentwicklung im algemeinen und die Beziehung zwischen dem Habitus von Eiskristallen und Depolarisierung von Licht im speziellen.

Warme, gemischtphasige und kalte Wolken wurden innerhald der AIDA Kammer produziert. Grösse und Habitus von Eiskristallen wurden mit bereits vorhandenen inorganischen Aerosolen, relativer Feuchte über Eis (RH_{ice}), Temperatur (T) und Experimentzeit von verschiedenen Experimenten in Beziehung gesetzt. In den Experimenten 1 und 3, welche mit unterkühltem Wasser beziehungsweise unterkühltem Wasser plus Schwefelsäure in der ersten Kampagne in Karlsruhe initiert wurden, dominierten reguläre Eiskristallformen. Die mit HOLIMO detektierten Hydrometeorgrössen befanden sich innerhalb von 2 und 140 μm (mit einem Durchschnittswert von 11 μm) beziehungsweise 2 und 50 μm (mit einem Durchschnittswert von 9 μm). Das 2. Experiment von derselben Kampagne wurde in einem ähnlichen T und RH_{ice} Bereich aber, im Gegensatz dazu, mit unterkühltem Wasser und Eiskeimen initiert. Die daraus resultierenden Eiskristalle wurden von irregulären Formen im Grössenbereich von 2 bis 168 μm dominiert (mit einem Durchschnittswert von 17 μm). Das 4. Experiment von dieser Kampagne wurde bei viel kälteren Temperaturen initiert, um den homogene Gefrierprozess von Schwefelsäure zu untersuchen. Innerhalb der Auflösung vom holographischen Mikroskop wurden nur Aggregate mit Grössen von 2 bis 43 μm (mit einem Durchschnittswert von 12 μm) gefunden. Speziell für Experiment 2 und für sehr dünne Plättchen wurden sehr kleine lineare Depolarisationen von bis zu 0.04 gefunden. Diese niedrigen Werte liegen unter allen bisher bekannten LIDAR Messungen für Wolkentröpfchen.

Mischphasen Wolken wurden auf der hochalpinen Forschungsstation Jungfraujoch untersucht. Die Frage war, ob sich in diesen Wolken Zonen mit nur Eiskristallen mit solchen reiner Wolkentropfen abwechseln. Die vorläufigen Resultate von unserem Experiment zeigen unterschiede zu den Häufigkeiten von Wassertröpfchen und Eiskristallen, die sich in einer gut durchmischten Wolk ergeben würde.

Chapter 1

Introduction

Large ice particles are abundant in Earth's atmosphere. Their radiative properties play a major role for Earth's climate because the radiative budget of the Earth is very sensitive to changes in cloud cover and habits of the ice crystals in the upper troposphere. There are indications that global cirrus coverage is increasing with time, probably due to increased air traffic ([1], [2], [3]) because aerosols and gases are directly emitted into the most sensitive region. Cirrus clouds are the dominant cloud type in the upper troposphere, covering approximately 20 to 30% of Earth's surface. They consist of ice crystals of various shapes in a broad size range from tens to several hundred μm. LIDAR depolarization measurements revealed simple shapes such as hexagonal plates and columns and more complex shaped crystals such as bullet rosettes and aggregates ([4], [5]). Ice crystals in clouds can freeze homogeneously or heterogeneously ([6]). In the heterogeneous pathway soot particles or mixed soot/sulfate particles emitted from air craft engines or mineral particles originating from the Earth surface may act as ice nuclei ([7], [8]). Much has been learned about ice nucleation and its importance in cloud microphysical processes ([9], [10], [11]). Nevertheless, there still is a lack of a comprehensive theory for heterogeneous ice nucleation ([12]). Therefore, it is necessary to obtain a better knowledge about the efficiency of aerosols to nucleate ice. Once nucleated the ice crystal growth will depend on relative humidity and temperature. Thus, different formation conditions will lead to different ice crystal habits. Those habits will also have different scattering properties and will depolarize light differently. Consequently, it is important to know the aerosol type, the conditions during possible ice formation and the ice crystal habit and orientation.

Cirrus cloud particles reflect a part of the incoming visible solar radiation back to space, which leads to a cooling of the atmosphere below (cloud albedo effect). On the other hand, they simultaneously decrease emission of terrestrial infrared radiation to space, which causes a warming of the atmosphere (green house effect). The properties of cirrus clouds like cloud cover, thickness and ice water content and the microphysical properties of the ice particles like size, shape and alignment determine if the cooling or warming dominates. From the perspective of particle optics, cloud solar albedo depends on the scattering cross section and the asymmetry of the angular scattering phase function which is the averaged scattered light over all scattering angles. In contrast, infrared warming depends on the infrared absorption properties of the crystals. In nature, for thin cirrus clouds, the shape of the ice crystals makes a big difference because mostly single scattering events take place. For thick cirrus clouds, multiple scattering events are predominant and average out the contribution of the shape to only 10%. In total, it is assumed that the warming from the decrease of terrestrial infrared radiation outweighs the cooling due to the solar albedo effect.

However, model calculations show that it depends strongly on the shape of the ice crystals. Thus, clouds composed of non-spherical crystals have a higher albedo than those composed of surface equivalent spherical crystals ([13]). The higher cloud albedo is thereby due to a lower forward scattering potential, predicted for non-spherical ice crystals by different optical models like geometric ray tracing. Additionally, cirrus clouds induced by anthropogenic emissions may be characterized by a larger number of small ice crystals. The strong backward scattering properties of these crystals may have a cooling effect on the atmosphere.

During the last decade, cirrus cloud nucleation and associated calculations of cloud radiative forcing have witnessed an increased amount of attention. [14] obtained a warming of the tropical tropopause of 1 to 2.5 K in January due to enhanced absorption of long wave radiation from the ground just by assuming planar polycrystals instead of columns or symmetrical spheres. [15] found that the shape of ice crystals is important for the amount of snow reaching the ground. If aggregates are assumed, then a 10-fold increase in aerosol concentration in the Arctic leads to an increase in accumulated snow by 40% after 7 hours of simulation whereas the snow amount decreases by 30% if planar crystals were used for the calculation. These studies show clearly that it is important to find a link between measurements of ice crystal habits and model calculations. Difficulties arose because the connections between ice nucleation and cloud properties are not well understood. [16] and [17] carried out model sensitivity studies about cloud evolution with respect to assumptions made for ice nucleation. Nevertheless, a link between good observations and parameterizations for ice nucleation is still missing.

It is also evident that there is a need for a closure between depolarization measurements from LIDAR and depolarization measurements at an ice nucleation chamber supported by imaging systems because there were difficulties to assign depolarization measurements of LIDAR to those of very thin plate like crystals. Usually, the values were so low that they were attributed to spherical particles. This issue can be addressed in laboratory experiments for instance when combining ice nucleation chambers with holographic instruments.

Holographic probes of different kinds are widely used in science and industry. Therefore, the technology of the various principles is very well understood owing to the pioneers that probably contributed most to this field ([18–199]). For instance, [96] used holography with electrons in order to investigate polymer structures. [162] investigated jets and sprays coming from a nozzle. Apart from that, also living organisms were widely studied ([96]). Even their 3D distribution and motion were investigated ([143]). Also in the field of Atmospheric and Climate Science holographic probes are used. In [30] and [54], for instance, they used a holographic probe for air borne measurements of hydrometeors. Others like [187] or [136] investigated atmospheric ice hydrometeors from the ground.

1.1 Ice crystal growth and habit formation

A motivation and also a guideline for this thesis is the ice crystal habit diagram adapted from [200] presented in figure 1.1. It shows four regimes (0 to -3°C, -3 to -10°C, -10 to -22°C and -22 to -35°C) of basic ice crystal habits inside a cloud with respect to temperature and complex structures with respect to ice supersaturation. The smallest ice crystals represented in this graph show sizes down to approximately 5 μm. The first and the third regime show plate like ice crystal habits. The second regime shows ice crystals with columnar and the fourth regime shows a mixture of plate and columnar ice crystal habits. This figure does not show the formation of the ice crystals. The supersaturation is given with respect to ice in gm^{-3}. The center point of the sketched ice crystals are taken for a calculation of the supersaturation with respect to ice in % using the Murphy-Koop parametrization for saturation vapor pressure above ice ([201]).

There is a gap in understanding the step from ice nucleation to final shapes even for pristine and simplified shapes outlined in figure 1.2. It is important to know under what circumstances ice nuclei will form certain habits and therefore, it is important to extend figure 1.1 to lower sizes.

Ice crystal growth and habit formation were intensively discussed over a big time pe-

Chapter 1. Introduction

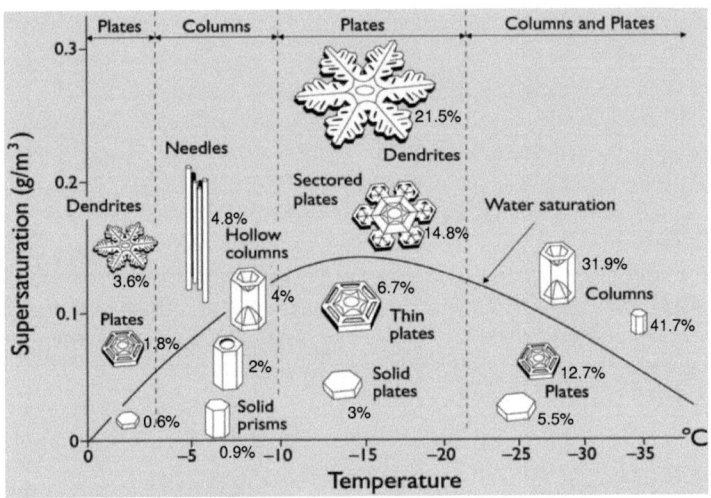

Figure 1.1: *Ice crystal habit diagram for in cloud crystal growth (adapted from [200]).*

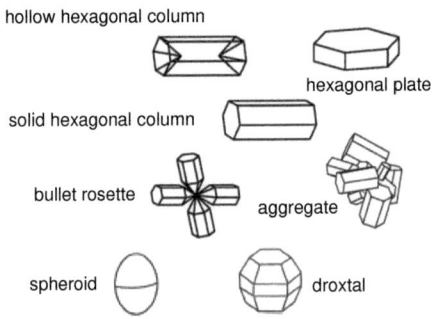

Figure 1.2: *Simplified shapes of possible ice crystals adapted from [202].*

1.1. Ice crystal growth and habit formation

riod in the scientific community. [203] and [204] discussed ice crystal growth and habit formation in great detail already 30 years ago. For instance, they studied linear growth rates of individual basal and prism faces of ice grown on a substrate as a function of temperature, excess vapor pressure and partial pressure of air. They found that effects of the environment (ice crystal growth in an environment of pure water vapor or grown at the presence of air) can be separated from surface kinetic energies. They also found that local maxima and minima of linear growth rates of basal and prism faces depend on the temperature characteristics of the primary habit. Figure 1.3 shows the different behavior of the different faces of a pristine ice crystal according to [205]. In one of their next studies they took the investigation one step further and made calculations on multiple layer formation of adsorbed water molecules on basal and prism faces of ice. They combined theoretical results with the outcome of a growth model in order to explain the observed strong dependence of, among other growth variables, the linear growth rate. They were able to explain the alteration of the primary habit with temperature (figure 1.4).

More recently, Libbrecht made calculations and measurements for the growth velocity normal to the surface of ice crystal growing from water vapor in terms of the Hertz-Knudsen formula ([206], [200] and [207]). He calculated the diffusional growth of ice crystals. He also found that existing growth data have been seriously distorted by systematic errors of one form or another. Hence there is a need for additional growth data

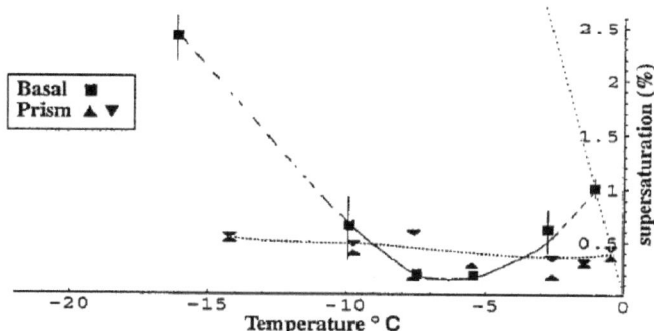

Figure 1.3: *Measured critical supersaturation with respect to ice and temperature at which growth was observed for the different faces of a pristine ice crystal ([205]).*

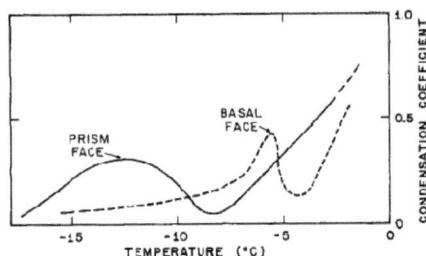

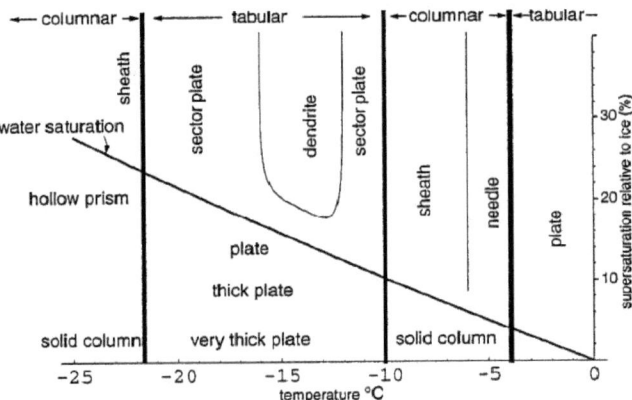

Figure 1.4: *Multi layer nucleation growth related to an ice crystal habit growth diagram. Combined date from [208] and [204]. The condensation coefficient on the top panel shows a sharp local maximum of the basal face in the columnar regime between -4 an -10 °C on the lower panel and it indicates the transition between sheath and needle like ice crystal habits. This local maximum in the upper panel is followed by a local minimum towards increasing* T. *It indicates the transition from columnar to tabular ice crystal habits at -4° C on the lower panel. The condensation coefficient of the prism face on the top panel shows a broad maximum between -10 and -15° C indicating dendritical growth inside the tabular regime on the lower panel between -10 and -22° C for supersaturations with respect to ice around 20% and more.*

and it should be treated with care in order to avoid such errors.

1.2 Outline of this thesis

In this thesis basic optical concepts and theory of operation of the HOLIMO (HOLographic Instrument for Microscopic Objects) detector and results obtained with this device will be presented in section 2, 3, 4, 5, 6, 7 and 8 respectively. More detailed, section 4 and 5 describe the important theory of digital holography and the performed proof of concept studies. Section 6 and 7 show nucleation chamber and field measurements respectively and section 8 contemplates briefly the ability of 3 D modelling of ice crystal habits from their holographic interference pattern.

A Holographic microscope was used because the principle of this technique allows for numerical treatment of 2D images with high and 3D shapes with reduced resolution of ice crystals throughout the whole observing volume V_{obs} provided that the information is recorded digitally, i.e. with a camera. That means that depth of focus, as it is crucial for optical microscopy, is not important for this system. This increases the amount of particles that can be detected during a measurement. In this thesis we show several times that with HOLIMO one can see objects smaller than 5 μm for instance some features of ice crystal particles presented in the result sections. It also has been shown that holographic probes can even resolve objects down to a size of 1 μm ([96]).

Chapter 2

Basic optical concepts of analog and digital holography

The next chapter describes basic concepts for optical systems on the basis of holography. Holography is a standard technology for measuring coherent interference pattern from objects with a light source and was first introduced by [56]. Gabor tried to come up with a proof of concept study for coherent electron beam interference considerations. He used a mercury lamp and narrowed down its emitting spectra in order to obtain a certain coherence length which then later was much easier to obtain in the early 60's after the invention of the laser. The name Holography was coined by Gabor. It is Greek for 'whole drawing' and means that in addition to the amplitude of the light coming from an object also its phase can be recorded. As an example of an application different from Atmospheric Science should be mentioned the production of forge proof items such as money.

2.1 Coherence length

Laser light can have a big coherence length cl whereas a LED, for example, only reveals partial coherence. In theory, this transition from no to any interference is very sharp. In practice, it depends on all the possible elements contained in the light path and whether the light source emits single or multi mode light. An upper limit is the cavity length of the laser and therefore, given by the manufacturer.

The importance of the cl can be best explained with the equations 2.1 and the in-line setup with diverging Point Source (PS) sketched on figure 2.2. In-line means that the light source interferes directly with the object. The object is situated inside a cone due

Chapter 2. Basic optical concepts of analog and digital holography

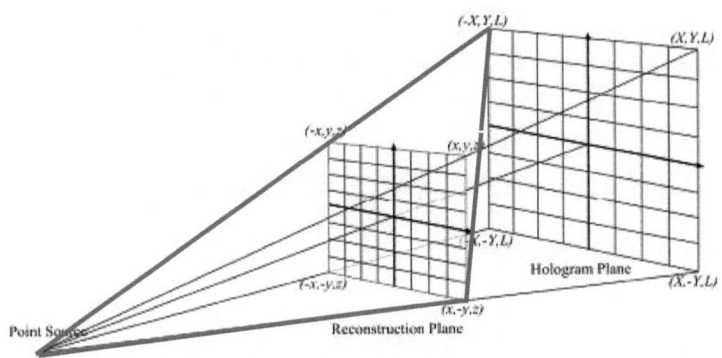

Figure 2.2: *Maximal path length difference (the cl indicated in blue) for an in-line holographic microscope for coherent interference imaging between the object and sensor plane. A point on the object and sensor plane is given by the coordinates (x,y,z) and (X,Y,Z) respectively.*

to the divergence of the PS. The interference pattern originates from the plane of the objects position producing a hologram on a screen from which the object can be reconstructed. Figure 2.2 shows the maximal possible path difference (cl) in such a setup. The coherence length can provide information about the interference of the amplitudes of the reference and the scattered waves. Equations 2.1 show the interference field of a single scatterer.

$$\left(\underbrace{\overbrace{l_i <}^{\text{coherent interference}}}_{\substack{(\vec{A}_{ref} + \vec{A}_{scat})^2 = \\ |\vec{A}_{ref}|^2 + \vec{A}_{ref}^* \vec{A}_{scat} + \vec{A}_{ref} \vec{A}_{scat}^* + |\vec{A}_{scat}|^2}} \quad \underbrace{cl}_{\text{ambivalent regime}} \quad \underbrace{< l_j}_{\substack{\text{absolute amplitude} \\ |\vec{A}_{ref}|^2 + |\vec{A}_{scat}|^2}} \right) \tag{2.1}$$

where l_i is the maximal path length difference and l_j is the resonator length. The light has a constant phase relation for all paths smaller than cl. It defines the cleanliness of the beam profile, the amount of modes and their linewidth. It is clear that only for $cl < l_i$ there will be phase functions to be recorded. Table 2.1 displays three examples for linewidths and corresponding coherence lengths. The first two examples can not be realized in a portable instrument. The last example shows a cl that is too little for our application.

Table 2.1: Linewidths and coherence lengths.

$\Delta\nu$	cl
3 MHz	31.8 m
90 MHz	1.1 m
30 GHz	3.2 mm

2.2 Depth of field and circle of confusion

The Depth Of Field (DOF) of a holographic instrument is its advantage over conventional microscopes. Figure 2.3 shows that all the points inside the DOF region will be displayed in focus inside region c of the image plane. It is called the circle of confusion plane of all points being displayed from within DOF. As a rule of thumb, c is 1/1500 of the focal length big ($\approx$11 μm in the case of the in-house fabricated HOLIMO (HOLographic Instrument for Microscopic Objects) detector, where the focal length can reach 16 mm). This Circle of Confusion (CoC) plane depends most naturally on the aperture of the setup. It can be decreased in order to increase the CoC but the contrast would also be decreased. It is not possible to obtain a big DOF for moving hydrometeors in conventional microscopy since either the exposure time is too big for imaging without blur, the intensity of the light source is so high that the objects could be damaged or the camera sensor could be overexposed.

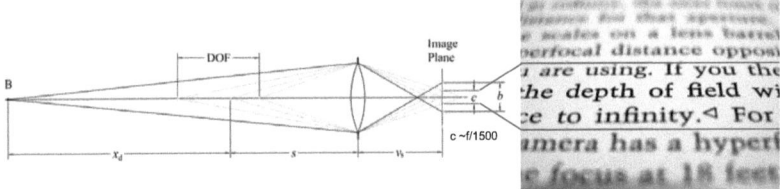

Figure 2.3: *The figure shows in the image plane parts of a tilted text in the object space illuminated by the PS at position **B** situated at distance x_d+**s** from the hologram. An object at distance **s** is in focus at image distance v_s. **b** indicates the region illuminated by the PS on the screen. Region **c** is the circle in the image plane where all the points lying within the DOF in the object room will be displayed in focus, hence the name Circle of Confusion plane. It is calculated from the focal length divided by 1500 as a rule of thumb ([231]).*

2.3 Aberrations and distortions

Aberrations express degrading sharpness image errors in the display of objects with respect to their position because an optical system can never reproduce the phase function of an object perfectly. This function can be expanded into a power series. Each power describes its own aberration. Figures 2.4, 2.5 and 2.6 show the aberrations with respect to the first, second, third and forth order of the object position in the power series expansion of the phase function respectively. The order of the figures represents therefore the strength of the effect on the image quality. Additionally, figure 2.7 and 2.8 show an error that influences the shape of an object only.

2.3. Aberrations and distortions

2.3.1 Astigmatic aberration

The astigmatic aberration (figure 2.4) is the inaptitude of representing points of lateral objects. Hence the structure representation of the objects can be an artefact.

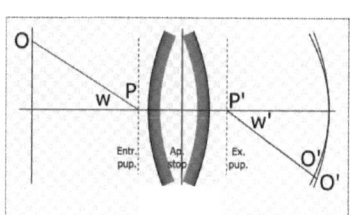

Figure 2.4: *The influence of astigmatic aberration on the image obtained with an optical instrument is shown on this figure. It goes with the object position with respect to the central axis to the power of 2 in the power series expansion of the phase function. The entrance into the pupil **P** of the ray **W** coming from the object **O** will be pictured from the exit of the pupil **P'** into **O'**. In this case there will be two **O'** because of the inaptitude of representing points of lateral objects. Characteristic astigmatic aberrations of a spherical object are shown on the right hand side ([231]).*

2.3.2 Coma aberration

The coma aberration (figure 2.5) is an asymmetry error of lateral objects. Hence, the bigger the asymmetry of an object the bigger the coma aberration which leads to a wrongly estimated shape.

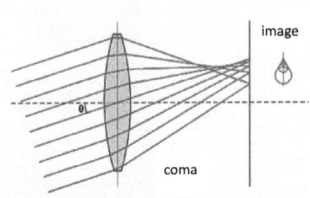 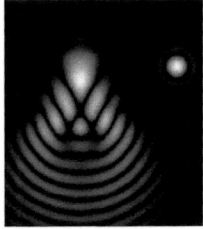

Figure 2.5: *The Coma aberration and its influence on the picture quality is shown on this figure. The error goes with the object position with respect to the central axis to the power of 3 in the power series expansion of the phase function. This lateral asymmetry is shown on the right hand for a spherical object. The small point would be a coma free representation of the object ([231]).*

2.3. Aberrations and distortions

2.3.3 Spherical aberration

The spherical aberration (figure 2.6) is an error of aperture and describes defocusing and blurring of an object with respect to its position towards the central axis. A PS, even being in focus, shows blurring in the third picture of the bottom row on figure 2.6. Negative and positive spherical aberrations are anti symmetrical with respect to the central image of the bottom row on figure 2.6 showing positive spherical aberration.

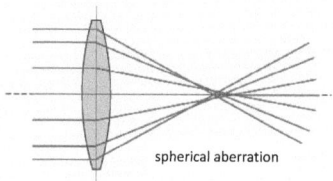

spherical aberration

Figure 2.6: *Affection of the image quality by spherical aberration sketched on top of the figure. It goes with the object position with respect to the central axis to the power of 4 in the power series expansion of the phase function. It is an error of aperture because of its finite extension. The row of pictures at the bottom of the figure shows a positive spherical aberration of a PS. Images to the left are defocused toward the inside, images on the right toward the outside ([231]).*

2.3.4 Distortions

There are also errors that do not degrade the sharpness of an object but its position. For instance, the whole image could show a pillow or barrel distortion (figure 2.7). This distortion could lead to a misinterpretation of the object shapes.

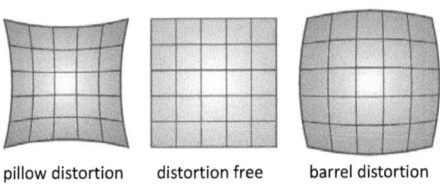

pillow distortion   distortion free   barrel distortion

Figure 2.7: *The possible distortions an optical instrument can show are sketched on this figure. They go with the object position with respect to the central axis to the first power in the power series expansion of the phase function. The left and right hand show pillow or barrel distortion of a lattice respectively. The lattice is displayed undistorted in the middle ([231]).*

Additionally, objects recorded far from the center of the camera sensor will appear bigger than those recorded in the middle because of the field curvature due to the divergence of the PS. They will have a different focal plane and hence a different inherent magnification. This introduces another source of error of object size, especially for small objects like the 5 μm PSL spheres on figure 2.8 obtained from one of the proof of concept studies discussed in chapter 5.

Finally, the kind of diffraction and the possible ways to get rid of them depends also in what diffraction regime the object is staying. This dependency is discussed in the subsequent chapter 3. [113] and [115] discussed in great detail multi order aberrations in holography and the consequences when they are transformed away.

The field of aberrations and distortions is an interesting topic but will not be further discussed in this thesis. Nevertheless, it should be mentioned that those errors could have been observed on some of the reconstructed images of hydrometeors obtained with HOLIMO for this thesis.

2.3. Aberrations and distortions 17

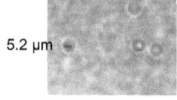

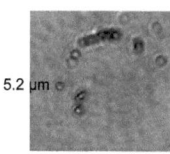

Parts of the same reconstruction

Figure 2.8: *The influence of the wave field curvature on the focal plane is shown on this figure. The picture on the top shows the part of the reconstruction of several 5 μm PSL spheres farther away from the center of the reconstruction plane than those on the lower picture. The focal planes of those PSL spheres will be different because of this lateral discrepancy between the first and the second group on the image.*

Chapter 3

Theory of operation of analog and digital holography

A holographic setup was chosen for this doctoral work because it offers a greater depth of field compared with conventional microscopy which can produce only one focal plane and therefore there will be a greater security and easier treatment in handling the objects position from a digital reconstruction of a hologram recorded with a camera. The only difference is that the phase of the light source is fixed over a certain distance and hence constraint. Reducing the parameters of light in real space offers a bigger variety in Fourier space. This means, in the case of data privacy like the one of figure 2.1, it would be more difficult to make a copy because of the 3 D nature of holographic images.

3.1 The Kirchhoff-Helmholtz (KH) transformation

Figure 3.1 points out the difference between the two regimes of Fresnel and Fraunhofer diffraction. The interference pattern of the object lies partly outside the shadow of aperture for the latter one. This means that the Fraunhofer diffraction pattern does not move as the object (together with the illuminating radiation) is moved laterally in contrast to the Fresnel diffraction pattern. This is also true for the propagation of the waves. The Fresnel diffraction patterns change as they propagate further away from the source of scattering, whereas the shape of the intensity of a Fraunhofer diffraction pattern stays constant. This is because Fraunhofer pattern is a function of angle only. It gets bigger the further away from the scatterer and its intensity smaller but the overall integrated intensity remains constant. The Fresnel number $N_f = \frac{a^2}{L\lambda}$ helps determine the regime based on a simple relationship between wavelength λ, object size a and the distance

L between light source and camera (figure 3.1). The interference pattern itself lives inside the fourier space hence it could be treated with a Fourier transformation in order to transform it into a real representation of the object that caused its appearance.

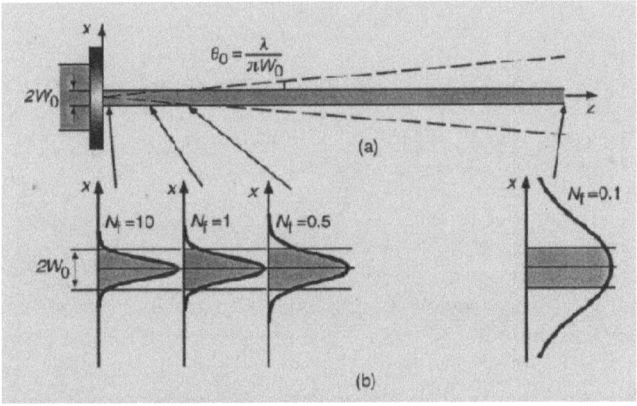

Figure 3.1: *This figure depicts the different diffraction regimes in (a). The Fresnel number N_f in (b) indicates the region of interest. Φ_0 gives the divergence angle of the laser and W_0 its extension at the exit ([232]).*

In fact, the interference pattern of a hologram is very similar to the time evolution of a Gaussian function in Fourier space. Therefore, if the unit in real space is m than the unit of the interference pattern would be m^{-1}. Normally, a hologram can not show the object without transformation but symmetric objects can reveal the same pattern and sometimes also the object can be seen, for instance, when a gauge in form of a ladder is used. This image only allows for measurements if the magnification of the imaging system would be 1 and hence the light source is a plane wave.

As a consequence, it would be best to express the transformation in terms of a Gaussian function or convolution because the calculation would be very easy. The next best thing would be an expression that is a direct Fourier transform in order to make use of existing and fast algorithms. Formula 3.1 shows the general expression used for transformation calculations in holography the so called Fresnel-Kirchhoff diffraction integral.

3.1. The Kirchhoff-Helmholtz (KH) transformation

$$U(P) = -\frac{iA}{2\lambda} \int\int_A \frac{e^{itk(r+s)}}{rs}[\cos(n,r) - \cos(n,s)]\mathrm{d}\xi. \tag{3.1}$$

Unfortunately, this is not a Gaussian and also not a direct Fourier transformation. Nevertheless, it is still possible to obtain a simple expression for the transformation when considering the Huygens-Fresnel principle of the convolution of elementary spherical waves that return the objects shape. It will look for the right points because it will be 0 all over except over the object (figure 3.2).

$$K(\vec{r}) = \int_A d^2\vec{\xi} \tilde{I} e^{\frac{2\pi i \vec{\xi} \vec{r}}{\lambda \xi}} \tag{3.2}$$

$$\vec{A}_{ref}(\vec{r}) = A_0 e^{i\vec{k}_{ref}\vec{r}}, \vec{A}_{scat}(\vec{r}) = A_0 e^{i\vec{k}_{scat}\vec{r}}$$

$$I = (\vec{A}_{ref} + \vec{A}_{scat})^2 = |\vec{A}_{ref}|^2 + \vec{A}_{ref}^* \vec{A}_{scat} + \vec{A}_{ref} \vec{A}_{scat}^* + |\vec{A}_{scat}|^2 \tag{3.3}$$

This is called the Kirchhoff-Helmholtz (KH) transformation for single or multiple scatterers (equation 3.2). Equations 3.3 and 3.4 and figure 3.3 show that multiple scatterers can be treated the same way as single scatterers. The wave from the light source is called reference wave and the diffracted waves from the objects are called scattered waves. All three waves will interfere coherently on the camera screen that records the interference pattern.

$$\begin{aligned} I &= (\vec{A}_{ref} + \vec{A}_{scat1} + \vec{A}_{scat2})^2 = |\vec{A}_{ref}|^2 + \vec{A}_{ref}^* \vec{A}_{scat1} + \vec{A}_{ref} \vec{A}_{scat1}^* \\ &+ \vec{A}_{ref}^* \vec{A}_{scat2} + \vec{A}_{ref} \vec{A}_{scat2}^* + \vec{A}_{scat1}^* \vec{A}_{scat2} \\ &+ \vec{A}_{scat1} \vec{A}_{scat2}^* + |\vec{A}_{scat1}|^2 + |\vec{A}_{scat2}|^2 \end{aligned} \tag{3.4}$$

The contribution of both scatterers alone is very weak compared to the reference wave and the interferences with it. What remains is the resulting interference amplitude between the scatterers and the reference wave and the background signal from it. The calculation is straight forward if there are more than two objects inside V_{obs}.

The Kirchhoff-Helmholtz transformation makes use of the Huygens-Fresnel principle. It looks for all the point sources within V_{obs}. Elementary waves are emerging from them. These sources are then weighted with their strengths $\tilde{I}$ and integrated over the whole

Chapter 3. Theory of operation of analog and digital holography

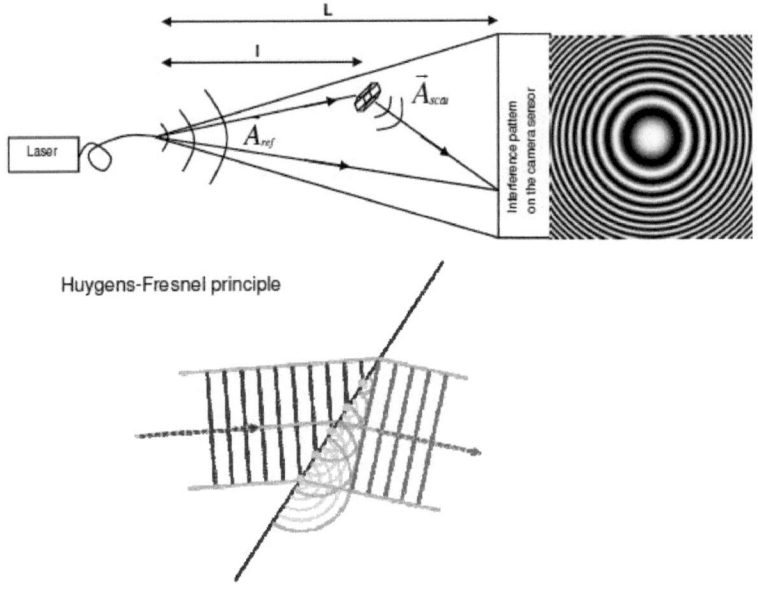

Figure 3.2: *This figure explains the Huygens-Fresnel principle for reconstructions from holograms. The upper sketch shows the in-line PS setup for a digital holographic microscope. The reference wave interferes directly with the object at distance l from the PS producing the interference pattern at distance L from the PS on a digital camera. The lower panel indicates the plane inside V_{obs} where the object will scatter the reference wave. Elementary waves are excited by a plane wave front (blue) at the yellow points of the surface of an object. A convolution of all the scatterers result in the reconstructed wave front (green) ([233]).*

recording area A in order to obtain the envelope wave front and hence a real representation of the object. As a consequence, the object needs to be at far field distances from both the PS of light and the camera. The far field condition $x_{crit} \geq (2a)^2 \lambda^{-1}$ takes the maximum diameter of an object $2a$ and the wavelength λ of the PS into account ([54]). For a spherical object with a diameter of 5 µm and with a wavelength of 532 nm x_{crit}

3.1. The Kirchhoff-Helmholtz (KH) transformation

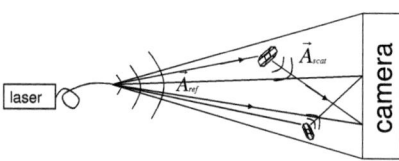

Figure 3.3: *Sketch of two scatterers inside the observing volume of HOLIMO.*

would be reached after 47 μm. The notion far field means that the wave front diverges although still being plane (figure 3.4).

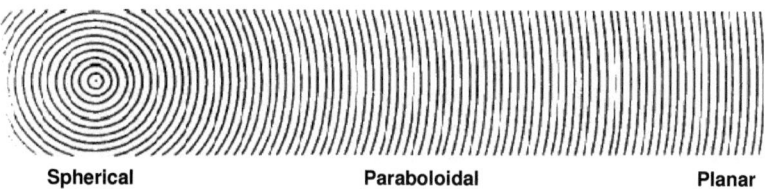

Spherical   **Paraboloidal**   **Planar**

Figure 3.4: *Wavefront behavior in the KH transformation. It takes place in the plane wavefront regime of the diverging wave ([232]).*

Paraxial approximation

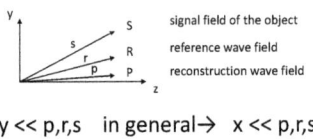

y << p,r,s   in general→   x << p,r,s

Figure 3.5: *This figure shows the paraxial approximation for in-line setups of optical instruments. The distance from a scatterer to the longitudinal z direction is a lot smaller than from the lateral y. The wave fields originating from the reference source, the object and the reconstruction source must have a small enough divergence. This means that the objects expansion in the y direction is a lot smaller than the distances p, r and s. If this holds then it holds in general also for the objects expansion in the x direction. The x axis is perpendicular to the y and z axes.*

Figures 3.5 and 3.6 show a basic assumption made in the KH transformation called the paraxial approximation. In the first place, every single spherical wavefront of the system should be considered but the expression of the KH transformation contains terms for plane wavefronts. This is due to the approximation made in equation 3.2. There we see that, by neglecting the smallest terms in the expression $\vec{\chi} - \vec{r}$, only terms with a vectorial dependence are left over. $\vec{\chi} - \vec{r}$ is the position of the object inside V_{obs}.

In figure 3.6 we can see that the position of a test point is given by $|\vec{\xi}| - |\vec{\xi} - \vec{r}|$ and because $|\vec{\xi} - \vec{r}| \approx |\vec{\xi}| - \frac{\vec{\xi}\vec{r}}{|\vec{\xi}|}$ this becomes $\frac{\vec{\xi}\vec{r}}{|\vec{\xi}|}$. Terms of higher orders like $\frac{|\vec{r}|^2}{|\vec{\xi}|^2}$ are very small if the angle between $\vec{\xi}$ and $\vec{r}$ is small and can therefore be neglected. Note that although the kernel is looking for spherical elementary waves the expression in the exponential becomes the one for plane elementary waves due to this approximation. Equation 3.2 will have two solutions. This can be seen easily because lens refraction and diffraction of a hologram can be treated analogous. The Newtonian thin lense equation $\frac{1}{f} = \frac{1}{z} - \frac{1}{L} \rightarrow z^2 + lz - lf \stackrel{!}{=} 0$ gives for $L = l + z$ 2 solutions if $l > 4f$ and hence the exponential in equation 3.2 is insensitive with respect to the sign of the phase. This is the twin image problem because the twin image will be superimposed on every reconstruction. The larger the object the weaker this contribution but the larger the influence of out of focus planes. Only large thin objects that are oriented perpendicular to the point of view are free from those effects. All possible image distortions described in section 2 can be present in a hologram because of this analogy.

3.2. Intensity distribution of the diffraction pattern

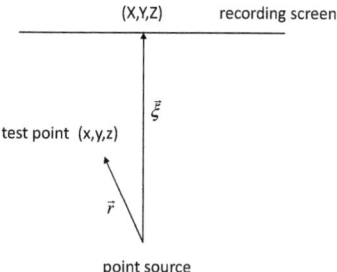

Figure 3.6: *This figure shows the wavefront approximation made for the KH transformation. The distance of an object at the test point (x,y,z) from the center point (X,Y,Z) on the screen is given by $\vec{\xi} - \vec{r}$. In the approximation this becomes $\frac{\vec{\xi}\vec{r}}{\xi}$.*

3.2 Intensity distribution of the diffraction pattern

The intensity distribution of a PS interfering coherently with a spherical object would result in a pattern sketched in figure 3.7.

This Bessel intensity distribution of the lobes and the first minima (m=1) at r_1 of $2\frac{J(g)}{g}$ can be calculated with the help of the following expressions:

$$
\begin{aligned}
r_1 &= r_{min} = 1.22\frac{\lambda_0|z_0|}{2a} = 78 \;\; \text{mm} \stackrel{!}{=} (1+m)Nd \to N \\
g &= \frac{2a\pi r}{\lambda|z_0|} \quad \text{and for} \quad J(g(r=r_{min})) = 0 \\
\to g &= 3.83 \quad \text{for} \quad 2a = 1 \;\; \mu\text{m}, z_0 = 12 \;\; \text{cm and } \lambda_0 = 532 \;\; \text{nm} \quad (3.5)
\end{aligned}
$$

The first minimum from an object recorded at a far-field distance of $Nd = N2a$ would be outside most commercial available films or camera sensors for this small particles. Only for particles which are a slightly larger than 10 μm the first minimum could be recorded with a standard quadratic camera sensor with a dimension of 15 mm. The distance to the next minima can be calculated with equation 3.6.

Chapter 3. Theory of operation of analog and digital holography

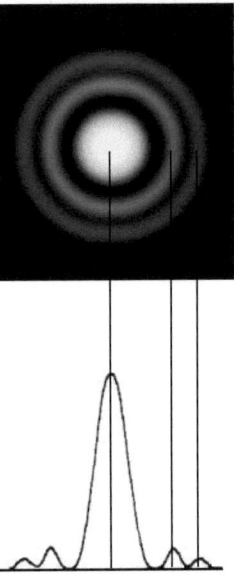

Figure 3.7: *Scattered intensity distribution of a spherical aperture. The central spot caries 84% of the overall light intensity ([234]).*

$$\begin{aligned} \Delta r(m) &= r_1(\sqrt{m} - \sqrt{m-1}) \\ r_n &= r_1 + \Delta r_2 + \Delta r_3 + ... \\ \Delta(m) &= \sqrt{m} - \sqrt{m-1} \to \frac{r_n}{r_1} - 1 = \sum_i = 2^n \Delta(i). \end{aligned} \quad (3.6)$$

If z_0 in equation 3.5 is shortened by 5 cm then already at object sizes of about 6 μm the first minima would be recorded with above described camera sensor as a worst case scenario. 84% of the light intensity is within the central spot of the Bessel intensity distribution.

3.3 Resolution of the camera

The camera needs to have a good resolution with respect to the amount of pixels and their sizes. It also has to have a good time resolution although, in imaging systems qualitative pictures are more important than quantitative results. In the end, a scientific grade CCD camera with 2048x2048 pixel with a pitch of 7.4 μm and a temporal resolution of 4 frames per second was chosen. This resolution is quite good among digital cameras but very week compared to a film. The test plate of AGFA 10E56 has got 4000 lines mm^{-1} which is equivalent to a 250 nm pixel size. The CCD camera has got a pixel size of 7.4 μm. This would meet with 135 lines mm^{-1}. So the plate has got a finer resolution but can only be used one time and needs by far more exposure time. Using a camera means making amplitude holography. A hologram is called a thick (3D) hologram, when its depth $d < \frac{2n\Lambda^2}{\lambda}$ where Λ and λ are the wavelengths of the reference and the reconstruction wave. In this case $\Lambda = \lambda$. So, for a silicon chip the pixel thickness d should not be $> 2n\lambda = 7\lambda = 3.724$ μm in order to be thin holograms. Because of the finite size of its pixels the camera has the tendency to show a skewed image due to the shape of the pixels. In this case, they are square shaped. Moreover, a CCD sensor can show blooming and smearing when overexposed (see figures 3.8 and 3.9) due to spill over from the pixels. The results are horizontal and vertical stripes in the picture. The resolution sampling criteria says that for m lines one wants to record there must be $2m$ lines recorded but this becomes only important when there is an apparent problem with undersampling described further in chapter 5. It also depends on the magnification of the object. This can be altered numerically by changing the pixel size in the reconstruction routine described in section E table E.1.

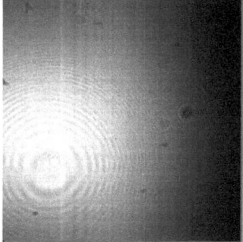

Figure 3.8: *Blooming and smearing of a hologram obtained with a CCD camera manifested in horizontal and vertical lines of overexposure.*

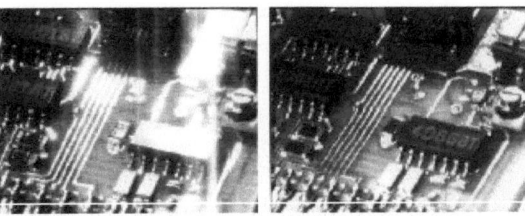

Figure 3.9: *Blooming and smearing of a picture obtained with a CCD camera for overexposure on the left and the same picture in normal exposure on the right operation ([231]).*

3.4 Energy density on the camera chip

The number of photons used for maximal and minimal levels can be estimated as follows. At 40MHz clocking rate the noise level would be $25e^-$ and the saturation signal would be reached with 20ke^-. Therefore, the minimum amount of photons would be $25e^-/QE$ and the maximum at $20ke^-/QE$ where QE stands for the quantum efficiency. At a clocking rate of 20MHz the noise level would be at $12e^-$ and the saturation at 40ke^-. If we want to express in energy vs. sensor area then $E_{min}/pixel$ and $E_{max}/pixel$ is going to be multiplied with the number of pixels per area. For a dynamical depth of 8, 10 or 12 bit the amount of pixels per gray level is 40ke^-/256, 1024 or 4096 giving 156, 39 or $10e^-$ respectively. The chosen sensor was a KODAK KAI 4021M sensor. It can be operated between -50 and $70°$, 5 and 90% RH and its $QE_{monochrome}(532\text{nm}) = 0.52$. The dynamical range is given by $20\log PNe/n_{e-T}$ with PNe the photo diode charge capacity and n_{e-T} the total noise. They depend on the date rate. At a clock of 20MHz PNe=40k and the noise is $12e^-$. Therefore, for these values the dynamical range becomes 71dB. The energy a photon is carrying is $E(\gamma) = h\nu = hc\lambda^{-1} = 3.73398 \cdot 10^{-19}$J. That results in a maximum energy density $\rho_{pixel}(E) = (E(\gamma)/A_{pixel})40ke^- = \frac{3.73\cdot 10^{-19}\text{J}}{54.76\cdot 10^{-12}m^2 QE(532\text{nm})} \cdot 40ke^- = 524 \cdot 10^{-6}Jm^{-2} \cdot e^-$. If we consider that the light cone is not equal to the CCD area, than $\rho(E) = \frac{CE \cdot E_{out}}{QE \cdot A_{light\_cone}} = \frac{0.6 \cdot 10^4}{0.52 \cdot 3.61}Jm^{-2} = 3.2 \cdot 10^3$ Jm^{-2}. The fiber coupler was adjusted in a way that only a single mode was observed. This was possible for a pulse energy below 1 μJ. The energy density of the incoming laser pulse was dimmed with neutral density filters because it was still too high for the camera sensor. Neutral density filters can be applied when the attenuation of the optical density OD after $OD = \log_{10}\frac{1}{T} \leftrightarrow T = 10^{-OD}$ is known. For $OD = 0.1 \rightarrow T = 0.794 \simeq 79.4\%$. In the end the pulse energy was set to 0.6 μJ and attenuated with 2 ND filters to 0.32 μJ. For pulse energies higher than 1 μJ more than just 1 mode was observed. These additional modes occurred probably due to Raman scattering (inelastic scattering) inside the fiber. The charge of an electron is $e = 1.602 \cdot 10^{-19}$C and with $Q = CU$ and $E = eV$ in [J] we can express everything in elemental charges.

The calculation without considering e^- noise or saturation signals at different clocking rates is a good first estimate. We have for different PS sizes different areas A and

$$A_{CCD} \to \rho_E = \frac{E_{in} \xrightarrow{CE} E_{out}}{A_{CCD}} \to \rho_E(e) = \rho_E Q_E$$

$$E_{in} = 1\mu J \to E_{out} = CEE_{in} = 0.6\mu J \to \rho_E(\gamma) = \frac{6}{3.61} \cdot 10^{-3} \text{Jm}^{-2}$$

$$\to \rho_E(e^-) = QE \cdot \frac{6}{3.61} \cdot 10^{-3} \text{Jm}^{-2} \tag{3.7}$$

Φ_{cone} =1 μm, 3.5μm or 5μm and therefore, the area A of the laser beam is $0.785, 9.621$ or $19.635 \cdot 10^{-12}$m^2 respectively. The area to be recorded at the sensor is $r_{diag} = \frac{1}{\sqrt{2}} \cdot 7.4$ μm$\cdot 2048 = 10.716 \cdot 10^{-3}m\to A_{cone\_cam} = r_{diag}^2 \pi = 3.61 \cdot 10^{-4}$m^2.

An example of the sensitivity range of a CCD camera is given by KODAK. It's between 10^{-7} and 10^{-4}Jm^{-2}. With a 1ns pulse this becomes 100 and 100000 Wm^{-2}. As a comparison the sunlight in winter and summer is between 200 and 1000 Wm^{-2}. The ray density of a collimated laser is calculated from $\phi = 1\mu$m$\to A = 0.785 \cdot 10^{-12}$m^2, $P = 3\mu$ J$\to S = P/A = 3.82 \cdot 10^6$Jm^{-2}. The couple efficiency is assumed to be between 30 and 50%. Fiber and fiber end will not impose a difference on single modularity and coherence length but on the polarization if the incident angle of light is different from 90°.

3.5 Light absorption of water and ice

For the application described in this thesis it is important to know whether the used light source power could melt or destroy the observed hydrometeors. The degree of absorbtion of water and ice with respect to a wide range of wavelengths is given in [209]. It can be seen that, for instance, ice absorbs good at wavelengths around 1500 nm but not at the 532 nm used in this work.

$$I(d) = I_0 e^{-ad} \to \frac{I(d)}{I_0} = e^{-ad} = \tau \text{ the transmission grade.} \qquad (3.8)$$

From [209] we get $a = \frac{4\pi k(\lambda)}{\lambda} = 0.3 \cdot 10^{-3}$ cm^{-1} for $\lambda = 532$ nm. k is the extinction coefficient. For a penetrating distances between 1 and 1000 μm, $\tau \simeq 100\%$. This means that ice will absorb almost no energy at a λ=532 nm.

After the first considerations were made a first version of the digital in-line holographic microscope had been constructed. This setup varied a little regarding the resolution during this doctoral work but in principle remained the same. Therefore, it is possible to investigate its basic properties in general.

Chapter 4

The holographic microscope HOLIMO

The setup of a possible digital in-line holographic microscope should not only be designed in combination with other instruments and nucleation chambers but also as a stand-alone instrument. Therefore, a simple setup with a laser as light source, an objective for the collimation, a pinhole to act as a PS and a camera to record the data was considered to be build. A laser, although not the only but best possibility, should be the perfect light source when coherent interference is required.

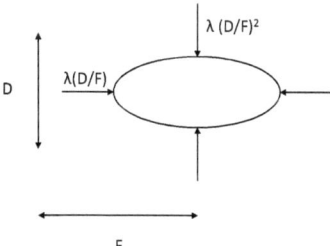

Figure 4.1: *Ability of focussing a laser beam in 3 dimensional space. D (laser source extension) is a lateral distance and F (distance between laser source and desired focal point) a longitudinal. A Gaussian beam can be collimated with D/F in lateral and $(D/F)^2$ in longitudinal distances.*

A laser possesses a very narrow power spectrum ν and small divergence of the emitted beam compared to other light sources like a LED. This means that the light emitted from the laser is cleaner than from any other light source having a comparable wavelength.

Moreover, a laser can be made for continuous wave and pulsed light applications. This is impossible for conventional microscopes for our application. Such a microscope can not have 5μm lateral resolution and a depth of field of several cm ([157]). If a resolution like this could be achieved with conventional microscopy than the exposure time would need to be very long. In fact it would be to long for measurements of moving objects. Thus, for moving objects a laser with a relatively short laser pulse is the right choice. This means that there is low fuzziness and therefore there is a certain divergence of the laser beam. This divergence differs between lateral and longitudinal directions. This is because a Gaussian laser beam can be better focussed in the first direction. More precisely, the lateral and the longitudinal ability to focus a laser beam are proportional to $\lambda \frac{D}{F}$ and $\lambda (\frac{D}{F})^2$ respectively. F is the distance of focussing from the light source and D is its beam diameter (figure 4.1).

4.1 Abblation due to deposited energy

The objective together with the pinhole can be used to produce a PS of light with a certain divergence and hence a certain inherent magnification. It can also ease the standards for the laser since its beam will be focused by the objective and cleaned by the pinhole. To forestall, the wavelength of the laser was chosen to be at a wavelength where the digital camera was most efficient in converting a photon into an electron. This was the case for $\lambda = 532$ nm. It would be best for the pinhole to be smaller than λ in order to have a true PS. This is already difficult since the laser beam needs to be collimated down to a beam waist that is only slightly bigger than the pinhole. There are two problems. The first one is illustrated in figure 4.2.

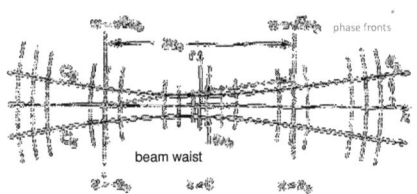

Figure 4.2: *Laser characteristics of a Gaussian beam.* r *indicates the point of focus of the laser beam.* $2w_0$ *is the point of minimal beam waist and* θ *gives the divergence angle of the beam ([232]).*

It can be seen that a laser beam can only be focussed down to a minimal beam waist

4.2. Camera laser trigger

in lateral and a different and larger one in longitudinal direction. The second one is the ability of a strongly collimated laser beam to ablate material if the area density is too large. The energy density is calculated from equation 4.1. For metal for $\lambda = 532$ nm it is $1 \cdot 10^4$ Jm^{-2} and for a 1ns pulse equal to $10 \cdot 10^{12}$ Wm^{-2}. For a non conductive material it is 5-15 times higher depending on the transparency of it. Air breakthrough, the point where air becomes conductive, would occur for densities higher than $50-100 \cdot 10^4$ Wm^{-2} which is a lot lower than the calculated energy density.

$$r_{min} = 1.17(\frac{f}{r})^2\lambda^2, \quad f = \text{focal length}, \quad r = \text{beam radius at origin}$$
$$P = \frac{E}{\tau}$$
$$L = \frac{P}{A_{min}} = \frac{P}{\pi r_{min}^2} \quad (4.1)$$

With equation 4.1 we see that ablation of any material would take place for r <133 μm. Therefore, the objective and the pinhole were replaced by a mono mode single frequency laser fiber. With this fiber a clean and coherent laser PS can be obtained with a beam diameter of 3.5 μm at the exit of the fiber.

4.2 Camera laser trigger

Although, HOLIMO consists merely of a camera and a laser it has to be decided how they are going to interact. This means that both need to be triggered as outlined in figure 4.3. There is a ramp of incertitude that comes with every trigger impulse but important is that the laser impulse is triggered in between an exposure period of the camera. The camera itself is quasi free running with its maximal time resolution of 4 frames per second. The framegrabber sets it to active and triggers the laser after 50% of the exposure period predefined in the camera setup.

Synchronization laser camera

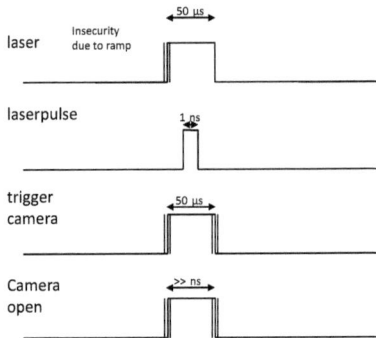

Figure 4.3: *Trigger considerations of HOLIMO. The camera and the laser need to be triggered at the same time. The laser pulse is initiated in the middle of the trigger signal. The camera needs to be exposed during a time period that covers the laser pulse.*

4.3 Ice crystal fall speed

It is important to know about the right setup for lab and field studies like the ones in Karlsruhe and on the Jungfraujoch. Several applications need to be considered. The design of the inlet might depend on the fall velocities. Ice crystals might orient themselves according to their habit inside the inlet system if the distance from the inlet entrance to V_{obs} of HOLIMO II is relatively long and the ambient air velocity inside the inlet system is relatively low i.e. when the ice crystals find the time to relax before getting measured. That means that flow speed and fall velocities need to be considered unless the crystals are falling directly into the small inlet of the instrument. Basically, HOLIMO II is not dependent on the flow rate. This has been shown at the AIDA chamber in Karlsruhe where flow rates from 5 to 160 lmin^{-1} were applied. Of course, somewhere in between this range the transition between laminar and turbulent flow will appear. Judging from the interference pattern this was roughly the case for a flow rate around 30 lmin^{-1} because then a chaotic behavior of the index of refraction could be observed. For an inlet of 4 mm inner diameter this would result in a flow speed of 40 ms^{-1} and a tube Reynolds number of 11888 at 0°C at norm pressure. The particle Reynolds number would be 9.5 at the same conditions for a particle diameter of 20 μm. The system can bear very low pressures since it was evacuated several times during the experiments at the AIDA chamber down to some hPa. To get hydrometeors into the inlet system the flow rate needs to balance the gravity force. By equating gravitational to aerodynamical drag force the following expression for the terminal velocity of a hydrometeor can be obtained:

$$v_T = \sqrt{\frac{2mg}{\rho A C_d}} \qquad (4.2)$$

with object mass, density of air, equivalent sphere surface and drag coefficient. The problem is that the drag coefficient $C_d = C_d(v_T)$ but it only becomes important for resinous fluid or creeping flow. Then it can be expressed via the Best and Reynolds numbers $X = \frac{2V_b(\rho_b - \rho_F)gD^2}{A\rho_F\nu^2}$ and $Re = \frac{vD}{\nu}$. In X there is the volume V_b, the density ρ_b, the cross section A and the maximal dimension D of the particle as well as the density ρ_F and kinematic viscosity ν of the fluid and the gravitational acceleration g. In Re there is the mean fluid velocity v and also the kinematic viscosity ν and the maximal dimension of the particle D. Eventually, $C_d = \frac{X}{Re^2} = \frac{2V_b(\rho_b - \rho_F)gD}{vA\rho_F\nu}$. The fall velocities of snowflakes are around 1-2 ms^{-1} and those of water droplets around 5-7 ms^{-1} ($\mu_{air}(0°C) = 17.4 \cdot 10^{-6}$ Pas, $\rho_{air}(0°C) = 1.293$ kgm^{-3}). Whether turbulent or laminar flow is required needs to be decided. Turbulent flow could orient hydrometeors randomly if the particles would otherwise have enough time to settle in the preferred orientation according to their habit. Such a flow could also lead to flow blocking inside the measuring cell. This could render a time evolved characterization of the hydrometeors nearly impossible. Such a flow blocking was observed during the HALO02 campaign in Karlsruhe. Ice crystals were gathering in one point of the measuring cell of HOLIMO II during such an event.

4.4 Application of the KH transformation

The distance from the PS to the camera divided by the distance from the PS to the object defines the geometrical magnification m_1. With a PS to camera distance of L=132.5 mm and a PS to object distance between 7.2 and 11.7 mm, m_1 is roughly between 11 and 18. This is true for point objects and spherical wave fronts in one medium. Otherwise, the magnification is changed as the wave front changes either in shape or divergence. The overall magnification is then $M = m_0 m_1$. For one medium

m_0 is given as the product $\lambda_{eff}^{-1}\lambda = n_1\lambda^{-1}\lambda = n_1$, where n_1 is the refractive index of and λ_{eff} the wavelength inside the window material. In our application

$$\begin{aligned}\lambda_{eff} &= \lambda m_0^{-1} = \lambda \frac{s_1 + s_2 + s_3 + s_4 + s_5}{n_1 s_1 + n_2 s_2 + n_3 s_3 + n_4 s_4 + n_5 s_5} \\ &= 532 \text{ nm} \cdot \frac{132.5 \text{ mm}}{140.8 \text{ mm}} = 501 \text{ nm},\end{aligned} \qquad (4.3)$$

where s_1, s_3, s_5 are the travelling distances of the laser beam in air and s_2 and s_4 are the travelling distances of the laser beam inside the windows with the appropriate refractive index $n_1 = n_3 = n_5 = 1$ and $n_2 = n_4 = 1.517$ [86]. The first window has a thickness of 6 mm and the second of 10 mm. This wavelength dependence becomes particularly important when a particle sits on a window. For a particle inside V_{obs} on the window closest to the PS $m_0 > 1$ because light eventually passes directly from the glass to the particle. Thus the particle appears to be 1.3 mm closer to the PS than it is and its size will be overestimated by 22%. On the contrary, the size of particles will be underestimated by 22% if they sit on the window closest to the camera inside V_{obs}. This wavelength dependency results in a sizing error that needs to be corrected for. A calibration with 20.2 μm PSL spheres yielded an average sizing error of HOLIMO of 15% for particles inside V_{obs} not touching any window. The interference pattern is treated with the Kirchhoff-Helmholtz transformation in order to obtain a real image of the object. This is appropriate because the Mie size parameter is much larger than 1 and therefore we are in the Kirchhoff regime (see below).

With HOLIMO a maximum amount of far fields $N = \lambda l M (2a)^{-2} = 2809$ would be possible between PS and object in the current setup. Such high values, owing to the magnification M, are ideal for intensity reasons. The overall intensity drops with increasing N values (see below) but the image to background intensity ratio increases ([182]). Equation 4.4 shows again the Kirchhoff-Helmholtz transformation and how it is used numerically with HOLIMO.

$$KH(\vec{r}) = \int_A d^2\vec{\xi} \tilde{I} e^{\frac{2\pi i \vec{\xi} \vec{r}}{\lambda \xi}}$$

$$KH(k,p;l) = \sum_{h=-n/2}^{n/2} \sum_{j=-n/2}^{n/2} \tilde{I}(h,j;L) e^{\frac{2\pi}{\lambda} \frac{\vec{k}h + \vec{p}j + lL}{\sqrt{h^2 + j^2 + L^2}}} \qquad (4.4)$$

K is one point of the reconstructed plane calculated at the point r inside V_{obs} and related to all the points ξ on the camera sensor by the light intensity $\tilde{I}$ at these points

4.4. Application of the KH transformation

at a fixed longitudinal distance L. Since the sensor of a camera has a finite amount of pixels equation 3.2 can be written as a double sum over n integers h and j. The expressions with tilde in the exponential represent distances in pixel size Δ_X and Δ_Y on the camera sensor and $\Delta_x = \Delta_X/M$ and $\Delta_y = \Delta_Y/M$ on the reconstructed image plane with respect to the center. For instance, $\tilde{p} = p\Delta_x$. K then is one point (k, p) of the reconstructed plane at a fixed longitudinal position l inside V_{obs} related to all the points (h, j) on the camera sensor by the light intensity $\tilde{I}$ at their positions at a fixed longitudinal distance L. Note that the integral expression of equation (4.4) does not determine the field of view of the reconstruction. Only when the size of x and y gets larger, i. e. the farther away the reconstruction plane is from the PS, a pyramid-like stack of images will be generated and a V_{obs} with respect to the Gaussian intensity distribution can be obtained.

If the transformation is calculated with the help of equation 4.4 then $n^2 \cdot n^2 = n^4$ terms need to calculated. For a $4x4$ matrix, there would be 256 terms. Hence, a sample calculation for a $2x2$ matrix with 16 terms is carried out in equation 4.5.

$$KH(k,p,l) = \sum_{h=1}^{n}\sum_{j=1}^{n} \tilde{I}(h,j;L)e^{\frac{2\pi i}{\lambda}\frac{\tilde{k}\tilde{h}+\tilde{p}\tilde{j}+rL}{\sqrt{\tilde{h}^2+\tilde{j}^2+L^2}}}$$

$$KH(-\Delta k/2, \Delta p/2, l) = \tilde{I}(-\Delta h/2, \Delta j/2; L)e^{\frac{2\pi i}{\lambda}\frac{\Delta k/2\Delta h/2 + \Delta p/2\Delta j/2 + lL}{\sqrt{(\Delta h/2)^2+(\Delta j/2)^2+L^2}}}$$
$$+ \tilde{I}(\Delta h/2, \Delta j/2; L)e^{\frac{2\pi i}{\lambda}\frac{-\Delta k/2\Delta h/2 + \Delta p/2\Delta j/2 + lL}{\sqrt{(\Delta h/2)^2+(\Delta j/2)^2+L^2}}}$$
$$+ \tilde{I}(-\Delta h/2, -\Delta j/2; L)e^{\frac{2\pi i}{\lambda}\frac{\Delta k/2\Delta h/2 - \Delta p/2\Delta j/2 + lL}{\sqrt{(\Delta h/2)^2+(\Delta j/2)^2+L^2}}}$$
$$+ \tilde{I}(\Delta h/2, -\Delta j/2; L)e^{\frac{2\pi i}{\lambda}\frac{-\Delta k/2\Delta h/2 - \Delta p/2\Delta j/2 + lL}{\sqrt{(\Delta h/2)^2+(\Delta j/2)^2+L^2}}}$$

(4.5)

and the same for the remaining 12 terms for $KH(-\Delta k/2, -\Delta p/2, r)$, $KH(\Delta k/2, \Delta p/2, r)$ and $KH(\Delta k/2, -\Delta p/2, r)$ with the appropriate kernel of the exponential function. It is possible to treat the nominator and the denominator, which represents the norm of the vector products of the kernel, differently. In this case there are 3 different norms in the 16 terms, namely $\sqrt{\frac{9}{2}\Delta^2 + r^2}$, $\sqrt{\frac{5}{2}\Delta^2 + r^2}$ and $\sqrt{\frac{1}{2}\Delta^2 + r^2}$. We see that the coefficient in front of the distance increment Δ is decreasing by about a factor of 2. If we look at the more general case of a $4x4$ matrix and if we assign a vector basis to the matrix field we can see that there are 16 fields and that the basis in array 4.6 is

$$\left(\begin{pmatrix} -3 \\ 3 \\ -3 \\ 1 \\ -3 \\ -1 \\ -3 \\ -3 \end{pmatrix} \begin{pmatrix} -1 \\ 3 \\ -1 \\ 1 \\ -1 \\ -1 \\ -1 \\ -3 \end{pmatrix} \begin{pmatrix} 1 \\ 3 \\ 1 \\ 1 \\ 1 \\ -1 \\ 1 \\ -3 \end{pmatrix} \begin{pmatrix} 3 \\ 3 \\ 3 \\ 1 \\ 3 \\ -1 \\ 3 \\ -3 \end{pmatrix} \right) \quad (4.6)$$

The first element (-3) from the upper left corner is called the pivot element. The matrix is build such that always 2 numbers from the column represent one field of the matrix.

In the case of a 4x4 matrix the pivot element is equal to $-(n-1)$. The next step shows that for the next element in the row the number 2 has to be added and for the next element in the column 2 has to be subtracted.

Using bases defined like this, every single detected object can be moved towards the center of the reconstruction by rotating the image of the interference pattern in multiples of 90° with a change in those bases. Additionally, for the determination of the maximal length of the object it can be turned in fragments of 90° within the plane of reconstruction. The biggest size obtained by them will be set to D_{max} from figure 4.12. The rotation of a square matrix by a fraction of 90° requires a rescaling of the matrix to its original size. This means that some part of the image matrix will be padded with artificial coefficients and introduce noise to the picture. Therefore, it is best to do this procedure after binarization. Then the coefficients can be set to background values (either 1 or 0). After binarization, the boundary box of the object is calculated by scanning the binarized image vertically and horizontally and accounting for every pixel that is different from the background. Especially, the first and the last object pixels from the vertical scan and the first and the last object pixels from the horizontal scan define the two sizes D_{max} and D_w from figure 4.12 and therefore the boundary box. The outer and inner circumference of the binarized object can be obtained by counting all the pixels that are seen first and last in a horizontal scan by checking whether the change is from background to object pixel or vice versa. Additionally, the apparent area of the binarized object can be obtained by counting all the pixels that are different from the background. Like this also holes would be allowed for. It is not advisable to correct for them because they could be real or artificial. Moreover, they can help distinguish droplets from spherical ice crystals, since droplets act like a lens and produce an artificial hole in the center of the spherical object. Not removing the holes can be a problem when looking at sphere equivalent diameters but not when looking at the maximal dimension of the object. For the experiments discussed in this thesis it was more important to look at the maximal dimension of the object since this is important for the interpretation of depolarization measurements.

4.5 Resolution considerations of the microscope

The first thing to chose is the resolution of the original hologram and hence also of the reconstruction. The maximal resolution of the camera used would be 2048x2048 pixels. It makes sense to bin the resolution down to 512x512 or even to 256x256 pixel resolution in some cases because $\frac{2048}{1024} \cdot \frac{2048}{1024} \cdot 2B = 8MB \overset{4\text{fps}}{\rightarrow} 32MB \cdot s^{-1}$ in full resolution. This binning also depends on magnification of the reconstruction plane and therefore also on the aperture of the microscope. Additionally to that, an original hologram has the full dynamical contrast levels of a 16 bit image because the 12 bit images obtained from the CCD camera account for the upper 12 bits. This means that, in case of the background removal, it is certain that there will be negative values. These values need to be cut off for the purpose of displaying the image. Therefore, the contrast is going to be depleted.

The resolution of a coherent imaging system with respect to the numerical aperture NA is one way to determine the resolution of HOLIMO. There is a lateral $d_{lat} = \lambda(NA)^{-1}$ and a longitudinal $d_{long} = \lambda(NA)^{-2}$ resolution ([67]). These describe the minimal size of an object to be represented truly after the Rayleigh criterion (figure 4.4).

There are also calculations that take the object position within V_{obs} into account ([86]). The resolution of a holographic instrument can be calculated very accurately but this need not lead necessarily to a more precise resolution limit. There are lots of other factors that influence the resolution of the system. For instance, there is the resolution of the camera sensor i.e. the amount of pixels ($N_{pix} = n \cdot n$). HOLIMO has a maximum $N_{pix} = 2048 \cdot 2048 = 4'194'304$ with a pixel size of 7.4 μm. If this resolution is too small the interference pattern will be undersampled and the image will contain a lot of noise in the best case ([183]). This sampling criteria is dependent on m_1. If m_1 is small then N_{pix} needs to be large. By far the most important point for the resolution consideration is that the camera needs to record as many interference fringes as possible. This leads to the counter intuitive statement that m_1 rather needs to be smaller than bigger because edge smearing will then be reduced. [210] calculated the relative edge smear w for a long wire of thickness $2a$ and he presented an analytical way to describe edge smearing

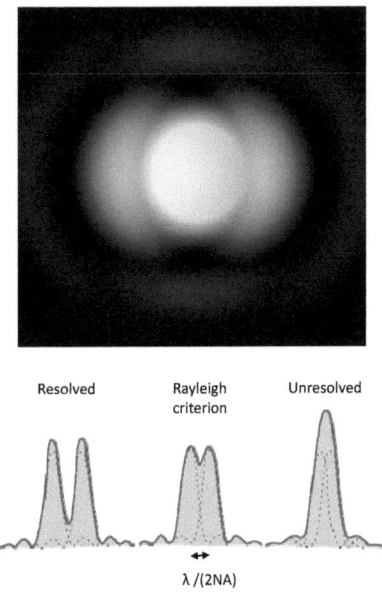

Figure 4.4: *This figure shows the Rayleigh resolution criterion for imaging systems. The top image shows the situation right in between the unresolved and the Rayleigh criterion case. They are sketched along the resolved case at the bottom of the figure ([234]).*

with respect to the amount of side lobes m of the interference pattern. Although derived for this special case he found the result $w(2a)^{-1} = (2m)^{-1}$ being valid for general cases as a worst case scenario. More quantitatively, particles with circular cross section will have a fringe spacing $\Delta r = 2a(1+m)^{-1}$ according to [33]. The maximal recordable radial distance is given by $r_{max} = (1+m)z\lambda M(2a)^{-1}$, with z being the distance between object and camera. Via conversion, the maximal recordable amount of interference fringes $m = 2ar_{max}(\lambda z M)^{-1} - 1$ can be obtained, with $r_{max} = 2048 \cdot 7.4$ μm=15 mm the maximal extension on the camera chip. This results in 9 fringes for a particle with a diameter of 5 μm, z =14 mm and for $M = 1$. The relative edge smear would then be 11%. Using spherical objects for resolution considerations is the worst case scenario because they reveal the worst resolution among all possible shapes. This is due to the curvature of their boundary. Therefore, they are more susceptible to noise as compared to a non-spherical object of the same size ([33]). In general it is agreed that $m = 3$ leads to a sharp enough image of the object ([33]). Additionally, spatial frequencies for

4.5. Resolution considerations of the microscope

coherent imaging $\nu = 2\pi r(\lambda z M)^{-1} = 8\pi(2a)^{-1}$ become independent of everything but the object size if $r = r_{max}$. Because the single mode single frequency laser pulse has a Gaussian intensity distribution the fringe visibility V for opaque objects goes like MN^{-1} ([180] and [183]). In fact, after [180] the visibility is

$$V(r) = -\frac{2Ke^{-\frac{r^2}{\omega^2}}}{K^2 + e^{-\frac{2r^2}{\omega^2}}},$$

whereas the factor K contains the normalized amplitude of the Bessel function divided by N, ω is the radius of the beam and r is the objects position in polar coordinates. This fringe visibility is bigger than with a uniform intensity distribution. It therefore makes sense, due to the complexity of the subject, to account only for λ and NA for resolution discussions. The numerical aperture of the used mono mode fibre in the far field is given by $NA_{eff} = 2\lambda(\pi MFD)^{-1}$ where MFD, the mode field diameter, equals 4.2 μm. This results in $NA_{eff} = 0.08$ which is equivalent to an aperture angle of 4.6°. An optical microscope with such an aperture angle would have a resolution limit of $d = \lambda(n \sin \alpha)^{-1} = \lambda(NA)^{-1}$ =6.6 μm after Abbe ([211]). It is the same as the lateral resolution of a coherent imaging system with respect to NA. As a consequence, for $M > 1.15$, the pixel size of the focal plane would be smaller than the resolution limit of diffraction for a pixel size of 7.4 μm of the recording plane. In other words, optical microscopes start to diffract objects of a size smaller than the Abbe limit and therefore, overestimate the size of the object. In general, holographic microscopes, having no constant magnification, can reproduce object sizes more reliably.

In a first step HOLIMO was optimized with respect to the maximum recordable area of the light cone which is defined as the area where the light intensity is higher or equal to the maximum intensity $I_0 \exp(-1)$. This means that the CCD chip (a square) is inscribed into the light cone from the divergent PS (see Figure 4.15). Whether the area of the camera sensor suffices for a clearly resolvable reconstruction of the object is hard to answer because it depends above all on the comparison of object sizes with the wavelength of the instrument. Mie theory gives an analytical expression for this comparison with its size parameter $(\pi 2a\lambda^{-1})$. If this size parameter is at least one order of magnitude bigger than 1 then there will be only forward scattered light within an angle of 3° ([54]). With this small divergence angle it is guaranteed that most of the scattered light intensity will be recorded with HOLIMO. For instance for a wavelength of 532 nm and a spherical object of 5 μm diameter this size parameter returns 29.5.

4.5.1 Edge blurring

Edge blurring is by far the most important factor when it comes to a degrading of the picture. This is because it contributes the most and it can be controlled to a certain degree. Figure 4.5 and 4.6 point out the problem that arises from edge blurring.

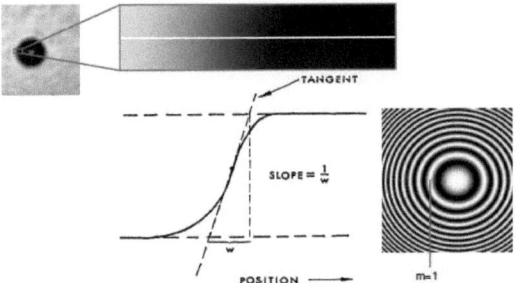

Figure 4.5: *The Influence of edge blurring on sharpness demonstrated on a droplet. The amount m of side lobes* m *of its interference pattern determines the slope $\frac{1}{w}$ of the transition from background to droplet ([210]).*

4.6. Particle hitting rate

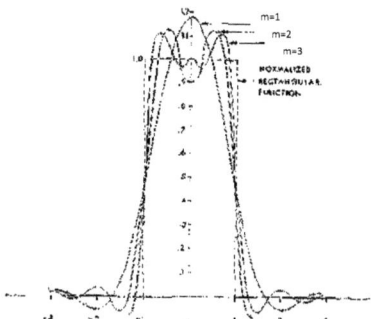

Figure 4.6: *The influence of edge blurring on a normalized rectangular function with respect to the amount of recorded interference fringes. All m reproduce the rectangle but the slope of the edges will increase with increasing m. Hence the size of the rectangle is more accurate with large m values ([210]).*

4.6 Particle hitting rate

HOLIMO has a V_{obs} of 8.3 mm^3 with respect to the Gaussian intensity distribution of the PS. The maximum speed of the camera is 4 frames per second. This means that HOLIMO is able to record $4V_{obs}$ per second. Therefore, the particle hitting rate is $c4V_{obs}$s^{-1}. This formula depends on the concentration c but it is independent of the particle flow. HOLIMO sees about 1 particle per 30 s if $c = 1 \cdot$ cm^{-3}. Nonetheless, it is important to deal with motion blurring if the velocity of the particle flow is too high. A rule of thumb is to allow a spherical particle to move 10 % of its size in the worst case. The laser used has a pulse length of 1ns. Therefore, for a flow of 10 l/min, the motion blur would be 53 nm which is less than 10% for a spherical object of 1 μm. Hence there is an upper limit to the flow concerning motion blur.

There is also a limit to the flow concerning laminar or turbulent flow regimes. This limit depends directly on the setup of the experiment. It is for certain that for a tube Reynolds number < 2300 there will be a laminar flow inside the tube.

There are two possibilities that can express the probability of seeing particles independent of the time because one considers all particles during an experiment and one compares their sizes with V_{obs}

$$p_1 = cV_{obs} \cdot fps \cdot t_{residence}$$
$$p_2 = 1 - \frac{4\pi}{3}\frac{(particle\_size)^3}{V_{obs}}$$
(4.7)

p_2 is important because V_{obs} is a cone and hence surface and volume of parts of the cone give different weight to the possibility of a particles position to be recorded.

$$V = \frac{\pi}{3}r^2 h$$
$$M = \pi r s$$
$$S = \pi r(r+s)$$
(4.8)

In equation 4.8 one sees that the volume and the surface of the cone of V_{obs} reacts quadratically and the mantel goes linearly with the objects position. It gets less probable for objects to be recorded closer to the PS because not only V_{obs} gets smaller but also the area were the particles can enter V_{obs} and the area were they can be recorded.

4.7 Data processing

In order to find particles at their correct position inside V_{obs} one has to find an algorithm to scan through it and eliminate unwanted/unfocused image planes. From now on this is being referred to as plane stacking. [162] found that coherent imaging has a big advantage over conventional imaging when it comes to the question of auto focussing. He found a way that was easy to apply noting that the average intensity increases towards the focal plane. Thus the particle can be found inside V_{obs}. This agrees with our findings (figure 4.7).

Figure 4.8 shows the brightness profile of an example. Panel A shows a relative brightness with a grey level separation of 263 and panel B of 126. The profile of panel B is very flat and therefore leads to a rejection of the position of this reconstruction plane.

After the brightest image has been found the maximum and the minimum of the average intensities of all reconstruction planes give the thresholds for the upper and lower limit for a segmentation resulting in a binary representation of the image. Thus the particle can be identified. Center of mass, outer and inner circumference, area and total area $\tilde{A}$ inside the outer circumference of the object are then determined from the binary picture. The center of mass can help to omit pixels that are wrongly attributed to the object due

4.7. Data processing

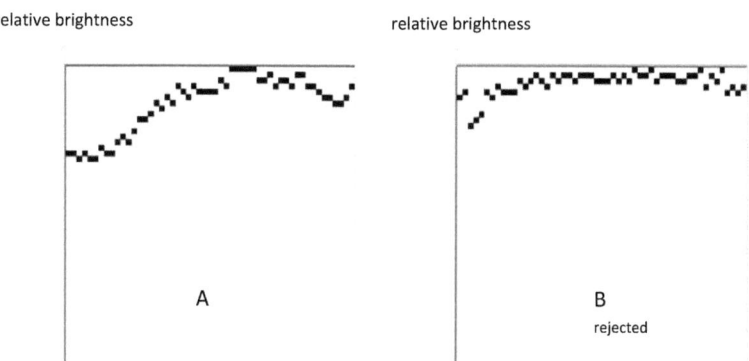

Figure 4.7: *Examples for the thresholding selection process. The 53 spots represent one reconstruction plane within V_{obs} between PS and camera. A flat histogram (the difference and outline between the spots given in relative brightness is smaller than within others) like the one at the right hand side leads to a rejection of the parameters attributed to this reconstruction plane when there is a histogram like the one at the left hand side within the reconstruction sequence.*

to the chosen threshold. This means that there is noise. If the noise is far away from the object then everything that is outside a radius that contains the object will be set to background. Finally, a boundary box is determined and a classification of the images of the particles can be done. Earlier works from the last century classified images of ice crystals in a great variety of classes and subclasses ([212]) by eye (figure 4.9).

[213] classified their data automatically into the 4 major habits for cloud particles, namely circular, dendrites/aggregates, columns/needles and irregulars (figure 4.10). They also found that classifying data in this way can be done most unambiguously. For HOLIMO only three classes were used namely droplikes (corresponding to circular by [213]), regulars and irregulars. Later discrimination of regulars into hexagonal and non-hexagonal shapes (includes needles and dendrites) and irregulars into pristine shapes with imperfections and aggregates where the individual pristine shapes are

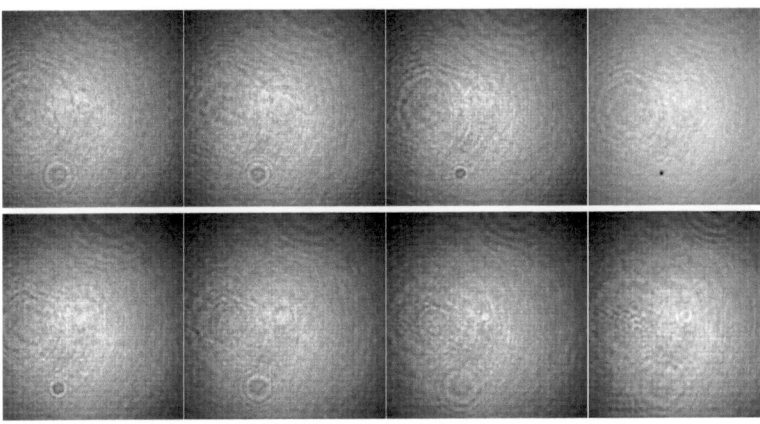

Figure 4.8: *Images illustrating the working principle of auto focussing. The reconstruction sequence starts at the top left image and ends at the bottom right. Somewhere in between V_{obs} there is one plane of reconstruction with the maximum mean brightness. It is the focal plane of the object.*

still visible was done by eye.

Figure 4.11 shows the automated process in a flow chart illustrated with an example. We decided to take the maximum dimensions in the x (D_w) and y (D_{max}) directions since the expected particles at the given atmospheric conditions for the experiment discussed here stem from the plate regime. Additionally, the area $\overline{A}$ of the object is determined. In this way, the aspect ratio $\alpha = D_w D_{max}^{-1}$, the roundness $\beta = 4\overline{A}(\pi D_{max}^2)^{-1}$ and the equivalent sphere diameter $d_{equiv} = 2\sqrt{\overline{A}\pi^{-1}}$ (it is used for comparison of round and regular shapes with $\alpha \approx 1$), allow for classification into the three classes mentioned above. The equivalent sphere diameter is represented as a binary sequence underneath the boundary box in figure 4.11. The first pixel is white and starts the binary sequence. It is not taken into account in the calculation. The amount of the leftover pixels -1 yields the highest exponent in this sequence and each white pixel indicates the appropriate coefficient to use. There is a sequence of 5 pixels indicating the objects size in the case shown on figure 4.11. This results in $D_w \approx d_{equiv} = (1\cdot 2^4 + 1\cdot 2^3 + 0\cdot 2^2 + 0\cdot 2^1 + 1\cdot 2^0)$ μm=25 μm. The outcome is used to assign the object to one of the classes with the help of the classification scheme illustrated in figure 4.12. This figure also shows α and β for different simple examples calculated analytically. Those values are giving a rather coarse

4.7. Data processing

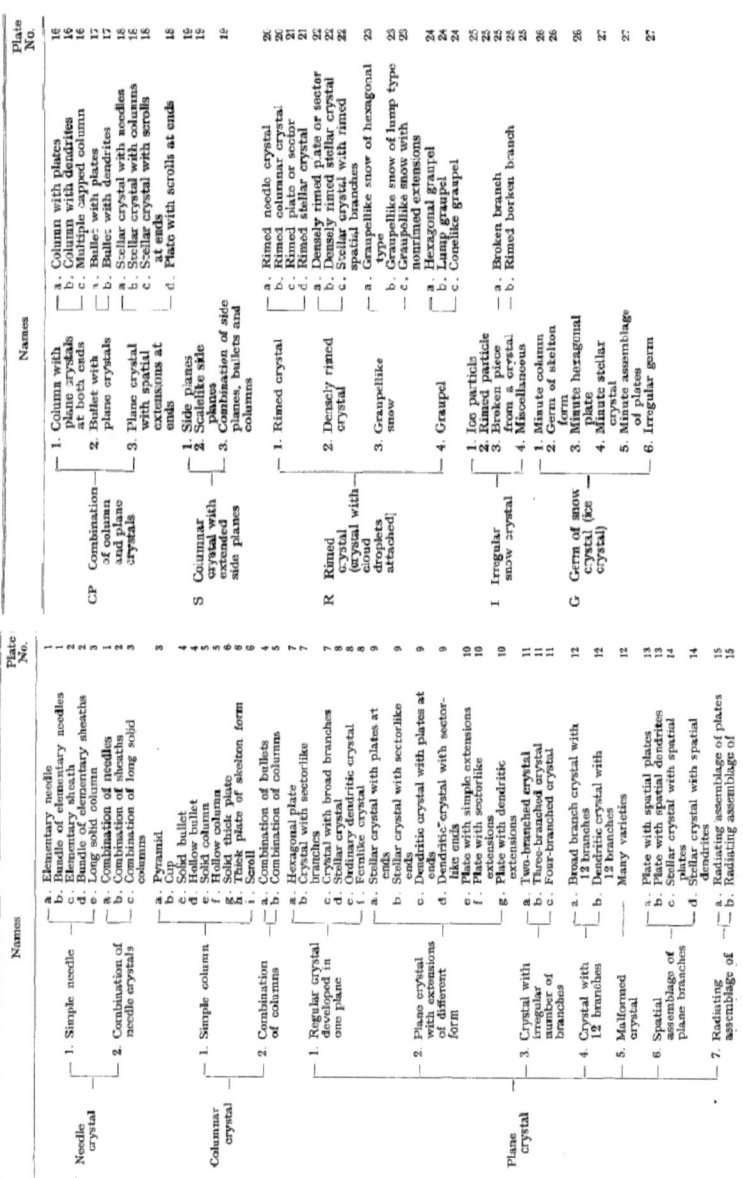

Figure 4.9: *Classification scheme by Magono et al. ([212])*

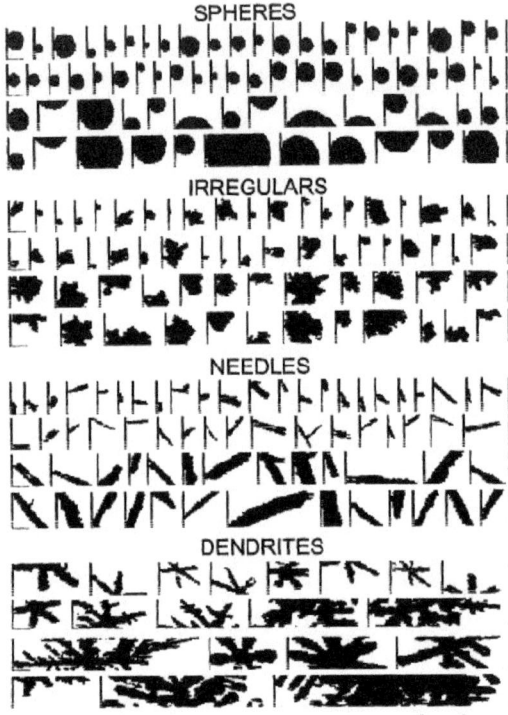

Figure 4.10: *Ice crystal classification after Korolev ([213]).*

constraint since an object is seldom recorded in the sketched orientation and there are no analytical expressions for more complex shapes. Therefore, values are determined from the inspection of possible ranges of such values for different experiments.

4.7. Data processing

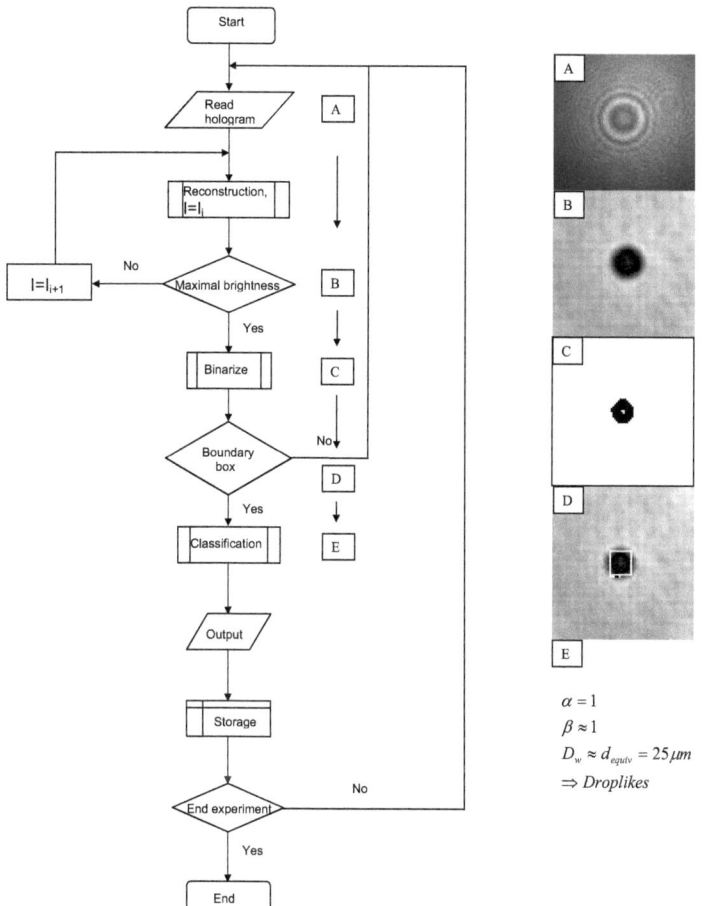

Figure 4.11: *Flow chart of the data processing of HOLIMO (left panel). First the hologram will be read in and a predefined routine reconstruction produces the image of maximal brightness at a distance l_j. The image needs to be binarized in order to define a boundary box. This makes it possible to classify the objects in a predefined routine and store the important findings. Every hologram is treated in the same manner before the data processing is ended. An example of this process is shown on the right hand side. Frame A shows the hologram, frame B its reconstruction and frame C its binary representation. Frame D includes the boundary box with the binary size label d_{equiv} underneath the box and frame E attributes the object to the class droplikes.*

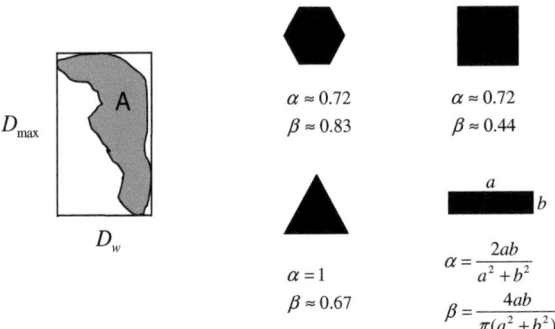

Figure 4.12: *Classification scheme for HOLIMO hydrometeor images. The measurable parameters D_w, D_{max}, A and the circumference are used for image habit recognition. The aspect ratio $\alpha = D_w D_{max}^{-1}$ and the roundness $\beta = 4A(\pi D_{max}^2)^{-1}$ are given for 4 simple shapes.*

There is also the possibility to classify the data with a clustering algorithm. This clustering algorithm was used to classify mass spectrometer data. Spectra can be produced out of object images via a change of variables from (x,y) to (r,φ) with subsequent Fourier transformation. This requires an additional fourier transformation and is therefore even more time consuming even though only the smallest section containing the object needs to be fourier transformed. The outcome of different classes might be too vast.

Figure 4.13 shows 20 different classes with respect to periodicities over an angle Φ of 360° for the first experiment from the IN11 campaign. Some images can clearly be combined to one background class but this has to be done by hand. Figure 4.13 also shows three panels of additional information. The color bar is showing the intensity of the class which is not used in this case. There is one panel indicating whether a particle from a class was found in an ascending RH_{ice} regime or a descending one. Another panel shows in what time period within the class a particle occurred and the last panel indicates the equivalent size of the objects found. This cluster analysis is a precious tool and gives lots of additional information but is laborious to handle and therefore, not used further on in this work.

Difficulties for the procedure of using an algorithm for the data processing are find-

4.7. Data processing

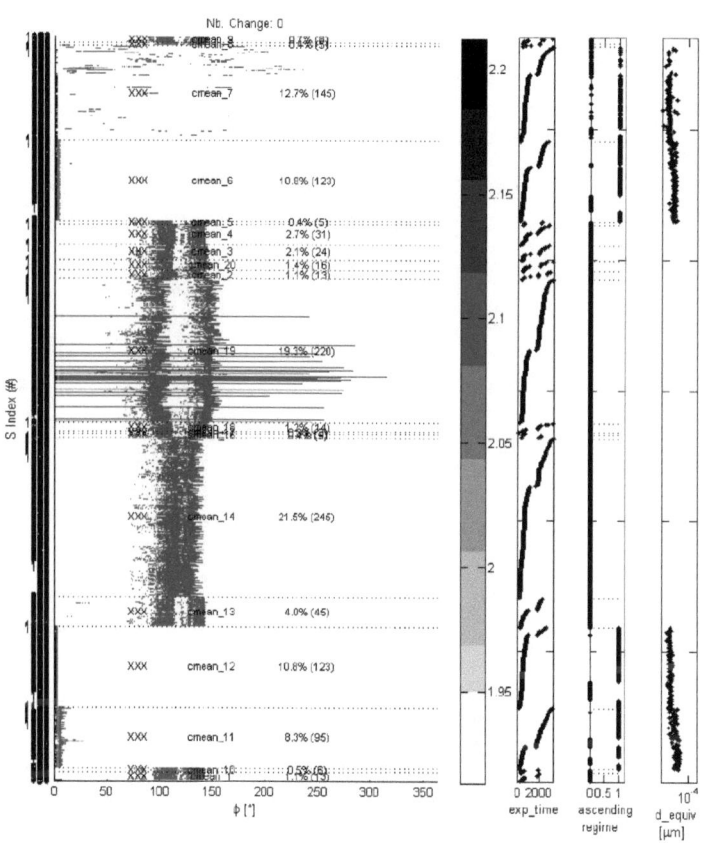

Figure 4.13: *Characterization of ice crystal habits with the help of a clustering algorithm. This figure shows 20 classes with different frequencies of occurrence. A lot of them are attributed to different backgrounds. This can be seen on the charts on the right hand side of the cluster classes. A size proxy from HOLIMO II shows the classes that are important for the characterization of the hydrometeors. The charts also show the time evolution of the hydrometeors inside each class and whether they were found in an ascending regime or not.*

ing a small enough reconstruction plane separation Δl, thresholding (introduction of background noise) and particles that are halfway over or fully outside of the Gaussian aperture. Particles that are too small or that occlude one another will be wrongly associated with their sizes. The habits maybe wrongly associated to a class when the reconstruction plane is not at the right focal distance. Therefore, the reconstruction plane separation needs to be decreased. That leads to an increased number of reconstruction planes inside V_{obs}. Nevertheless, increasing the number of planes to be stacked leads to an increased computing time. Therefore, a trade off must be made. Shapes of particles smaller than 10 μm can not be resolved if m_1 and/or m are too small.

4.8 HOLIMO I setup

The setup of HOLIMO I was realized 1:1 after the sketch of figure 4.14. A glass slide can be fixed on a rake at several distances from the laser light source and the camera. The chamber carrying the rake can be rotated for the purpose of easy manipulating the glass slides.

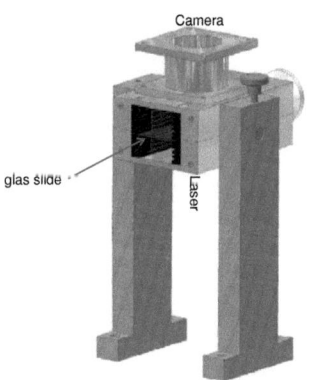

Figure 4.14: *Figure drawing of HOLIMO I.*

4.9 HOLIMO II setup

HOLIMO II is a Point Source Digital In-line Holographic Microscope and used to measure a sample flow. Figure 4.15 shows a sketch of the working principle of this instrument and an example image of interference fringes as it can be observed on a screen measuring a flow. A laser pulse with a pulse length of 1 ns, a pulse energy of 0.32 μJ and a wavelength of 532 nm is coupled into a single mode fiber of 3.5 μm core diameter. The fiber end acts as a PS of light. Eventually, the coherent light is scattered off from an object inside the light cone, defined by the divergence of the PS. This object is sucked through a sample flow tube with a diameter of 4 mm. The resulting coherent interference pattern is then recorded with a digital camera sensor. The camera (SVS4021 12BIT-S/VISTEK) and the laser (FDDS532-Q2/CryLas) are sealed off from the sample

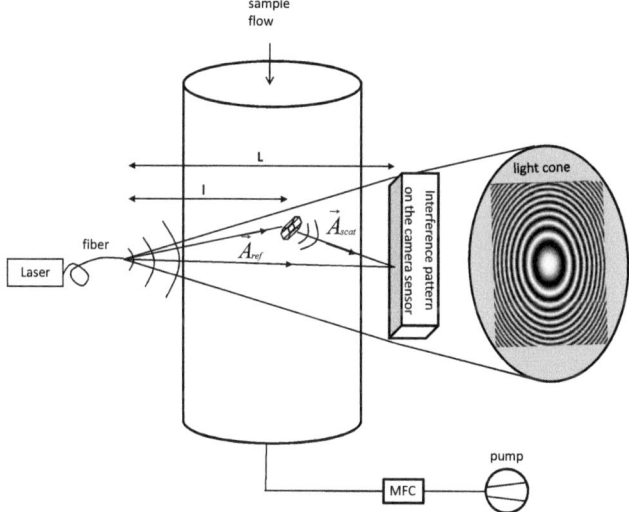

Figure 4.15: *Sketch of the working principle of HOLIMO. It shows the recording setup. The reconstruction is done numerically. Basically the setup consists of a laser PS of light and a CCD camera that records the interference pattern of the reference wave with the scattered wave amplitude from an object inside the sample flow tube. A possible interference pattern on the camera sensor inside the light cone is shown on the right hand side of the sketch. Objects are sucked through the sample flow tube with the help of a vacuum pump. The mass flow controller (MFC) controls the flow.*

flow with windows (not shown). The single mode fiber ensures that the laser pulse is a single mode single frequency pulse. As such the spectral width of the laser will be proportional to its coherence length $c(\pi\nu)^{-1}$ which is the relative travelling distance between the undisturbed reference wave and the scattered wave amplitude. Hence, it is a spatial and a temporal coherence length. The laser used in HOLIMO has a coherence length of 5 cm. The divergence of the single mode fiber and the setup of HOLIMO defines a light cone with a V_{obs} of about 8.3 mm^3. The detector as the base and the point source as the tip define a pyramid shape where V_{obs} is the part confined between the two windows. The exact volume is hard to determine because, due to the Gaussian intensity distribution of the PS, there is also the possibility that objects outside the light cone can scatter light back towards the camera.

Accordingly, HOLIMO consists only of a laser, an optical fiber, a camera, two windows and a small chamber confining the particle flow. With this simple setup one is able to determine size, position, orientation and shape of objects coming from a sample inlet. The windows were chosen after the following criteria

$$\begin{pmatrix} \text{ISO} & \text{Milspecs} & \text{comment} \\ \text{mm}^2 & \text{scratch-dig number} & \\ & 80-40 & \text{bad} \\ 0.25 & 40-20 & \\ 2x0.1 & 20-10 & \\ 3x0.063 & 5-10 & \text{excellent} \end{pmatrix}. \qquad (4.9)$$

Scratches (any marking or tearing of the surface) and digs (small rough spots and bubbles) are determined by eye. The first column gives the ISO values in mm^2, the second shows the scratch-dig numbers (how many can be seen), so called Milspecs, and the last column gives an indication on how good the windows are for optical applications. For example, when considering interference pattern then 5-10 Milspecs are advisable. The complex refractive index $n = n_1 + i \cdot n_2$ where $Im(n) = n_1 =$ absorbtion and $Re(n) = n_2 = \frac{c}{c_0} = 3.5$ for silicon. The refractive index of two glas species were considered ($n_2 = n_{borosilicate} = 1.517, n_2 = n_{fused\_silica} = 1.46$). The Anti Reflex AR coating for windows is safe until 10^4 Jm^{-2} for an energy of 0.33 μJ and a laser beam diameter of 1.75 μm $\rightarrow A = 9.621 \cdot 10^{-12}$ m^2 the power would be $3.326 \cdot 10^4$ Jm^{-2} that would be critical.

For the application in cold environments like the AIDA chamber or the Jungfraujoch two thermostate housings were build. The first one consisted only of thermal insulation plates shown in figure 4.16. Two boxes were made, one for the laser and one for the camera. Inside each box a selfregulating heater with 60 W power at 20°C was

4.9. HOLIMO II setup

positioned. In order to know how big every box needed to be the following calculation was used for the purpose of calculating thermal power loss.

$$\dot{Q}[\omega] = \frac{\lambda_D}{l} A \Delta T \qquad (4.10)$$

λ_D is the thermal conductance, A the area of the insulation material, l is the length of the ashlar and ΔT the temperature difference of desired temperature inside the boxes and the temperature of the sample flow. The insulation material used had a thermal conductance of 0.024 Wm^{-1}K^{-1}. $\dot{Q}[\omega] = 5.2$ W for a $A = 6 \cdot 1/6$ m^2, $l = 6 \cdot 23$ cm and a $\Delta T = 50$ K. The laser was tested for operating conditions between -5 and $35°$C and it is guaranteed to work at low pressure like 800 hPa because it is pressure sealed. The camera was also tested for operating temperatures between -5 and $35°$C. The camera, laser and its power supply plus the two selfregulating heaters were placed inside the second thermostate housing seen on figure 4.17.

Figures 4.16 and 4.17 show the setup of HOLIMO II at the AIDA chamber in December 2007 and 2008.

Figure 4.16: *HOLIMO II at the AIDA facility in November 2007. Camera and laser are distributed to two boxes made out of thermal insulating material. The red arrows indicate the position of the camera, the laser, its power supply and the heaters. The measuring cell (not shown) would be attached to the camera on the other side of the insulating material. The laser is about 28 cm long.*

Figure 4.17: *Setup of HOLIMO II at the AIDA chamber in December 2008. The measuring cell, the camera, the laser, its power supply and the heaters are placed inside the aluminum box. There position is indicated with red arrows.*

The main parts of the HOLIMO instrument are the laser and the camera. This simple two component system proofed to be adaptable very easy for different measuring opportunities like the laboratory and field measurements shown in the following three chapters.

Chapter 5

Proof of concept studies

The first tests were done with HOLIMO I (figure 4.14). This lab instrument can be used to investigate objects on a standard glas slide for conventional microscopes. The first thing to be thoroughly investigated with an optical instrument is a resolution target.

5.1 USAF target for HOLIMO I

It was decided to use the USAF1951 (US Air Force) resolution target (figure 5.1). It shows a descending helix order with smaller and smaller pairs of horizontal and vertical elements. Those pairs are black and void lines. Equation 5.1 shows how the resolution can be calculated with the help of this target.

$$f = 2^{(G+\frac{E-1}{6})} \qquad (5.1)$$

G stands for the group number and E for the element in this group. The resolution is given to distinguish object details in lmm^{-1}. The contrast is given as the maximum difference of bright and dark areas of an image given in % and the Modulation Transfer Function measures the ability to transfer contrast from object to image plane. The DOF determines how good resolved objects appear when they are out of focus. Distortion describes optical errors in %. The maximal resolution is therefore $2^{7+\frac{6-1}{6}} lmm^{-1}$=228 lmm^{-1}. In other words, the smallest line to see on this target is about 4.4 μm.

It is important to determine the exact position of the glas slide within V_{obs} and hence the magnification from the instrument. Since holography deals with interference pattern it is not obvious which setup is the best with respect to the resolution for the parameters that describe the distance between PS and CCD camera sensor and PS and object (L and l) and therefore also the magnification $m_1 = \frac{L}{l}$. This question is approached by looking at different parameter pairs (L, l). First, the possible range of magnification is fixed and the two extreme values are set to take the image of the target. Second, the range is decreased.

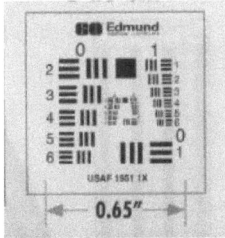 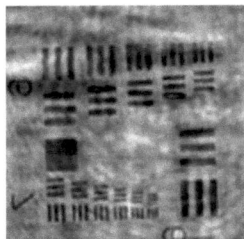

Figure 5.1: *The UASF1951 target shows a descending helix of horizontal and vertical pairs of lines in groups and elements. The resolution limit for the highest pair of group and element number (7,6) is equal to about 4.4 µm. This highest group and element number pair can be identified on the reconstruction of the target to its right for a PS to CCD distance of 8 cm and a geometrical magnification of 9 (target and target picture on the left from [235]).*

Third, the resolution of the original hologram was investigated for the best case setup. The PS to CCD distance was set to 13, 10 and 8 cm. This lead to a geometrical magnification between 4 and 16, 3 and 12 and 2 and 9. The resolution obtained for these three cases were 7.0, 5.5 and 4.4 µm respectively. Table 5.1 summarizes those findings.

Table 5.1: Geometrical magnification vs. resolution.

range of geometrical magnification	resolution
4-16	7.0 µm
3-12	5.5 µm
2-9	4.4 µm

It is very interesting that, apparently, the image has got a better resolution for smaller M even within a specific range of L and its ratio with l. This is the uncertainty principle of

5.2. PSL targets for HOLIMO I

holography: the closer the object to the PS the more uncertain its position or the more fuzzy the object. Forth, when the resolution is increased in the reconstruction by decreasing M than the resolution of the original hologram needs to be increased as well. This is because of the undersampling introduced by smaller M. This is the sampling theorem for holograms: twice as many interference fringes need to be recorded than the lowest resolution could provide in order to not undersample a hologram i.e. avoiding noisy pictures.

5.2 PSL targets for HOLIMO I

After this characterization of the instrument test images of 5.01 ± 0.035 μm PSL spheres with a standard deviation of 0.19 μm on glass slides were taken in order to see whether the resolution would be better, worse or equal to the one obtained with the target. ± 0.035 μm is the uncertainty in measurement from the manufacturer. Size determination was done for some particles grouped in figure 5.2 with HOLIMO. This figure revealed that the lower limit is about 4.4 μm for the group of PSL spheres on image *(b)* for the setup chosen. The indicated example of the group of PSL spheres on image *(a)* revealed an error of more than 50% which is a value far beyond the given range of ± 0.19 μm. One can see that the edges of the PSL spheres on image *(a)* appear to be very thick compared to the ones on image *(b)*. This makes it nearly impossible to quantify measurements on such small objects. There are considerations concerning edge blurring and other aspects discussed elsewhere in this thesis. Also the way preparing the glass slide can introduce large errors due to the low quality of it. Later on, an improvement of the lower resolution limit was made for one campaign in 2008 and one in 2009 for HOLIMO II. HOLIMO II is the instrument capable to measure a sample flow V_{obs} was changed due to resolution increase. The distance PS to CCD was decreased. This lead to an increase of V_{obs} from 8.3 mm^3 to 109.6 mm^3.

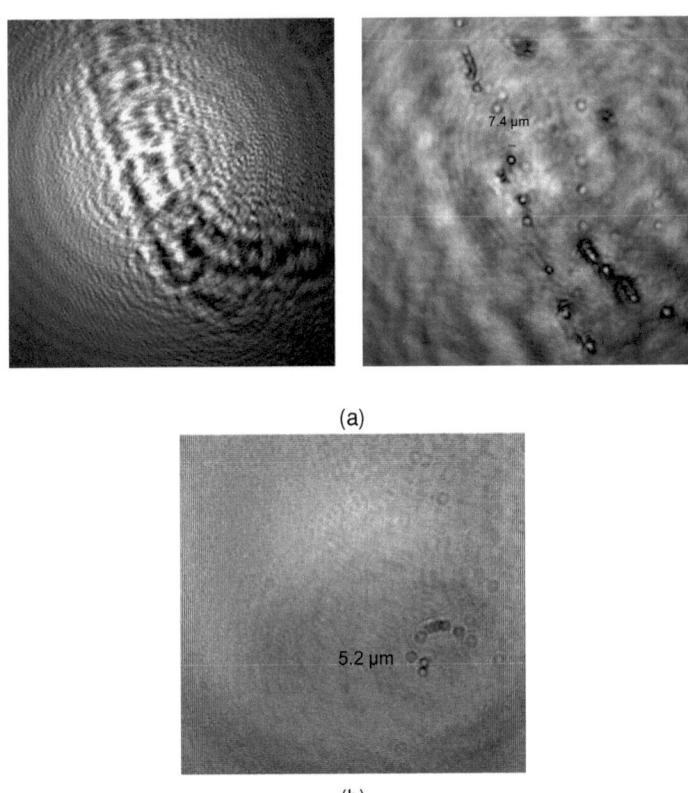

Figure 5.2: *Different reconstruction of different 5 µm PSL spheres resulting in different resolutions of them. (a) shows the interference pattern of a group of 5 µm PSL spheres on the left, the distribution of the spheres is recognizable, and their reconstruction on the right. They show a thicker rim of the 5 µm PSL spheres than the reconstructed group of PSL spheres shown in (b). This causes the resolution to differ about 50% from 5 µm probably due to edge blurring.*

5.3 Ice analogue targets for HOLIMO I

On figure 5.3 images of ice analogues ([214]) can be seen. These analogues are small microscopic crystals made out of sodium hexa fluorosilicate (Na_2SiF_6) with the

5.3. Ice analogue targets for HOLIMO I

same hexagonal plate and column like structure like ice crystals. They can also form out more complex structures like rosettes and their index of refraction of 1.31 is very similar to that of ice. They are stable under ambient conditions and therefore suitable for test measurements. They are fixed on plastic microscope slides covered with standard optical glass slides with an index of refraction of 1.49. It was unclear whether the sample holders are suitable for holographic imaging. Figure 5.2 shows that in fact the sample holders are not perfect for holographic imaging since there is a lot of noise in the pictures and the contrast is not optimal as well but, nevertheless, decent reconstructions could have been made. The sizes seen with HOLIMO I differ about 10% compared to the sizes measured by [214] in the worst case.

The first image in the top left corner on figure 5.3 (a) shows a SEM image from a analogue made by [214]. From this image moving clockwise towards an image from the same object taken with HOLIMO I and subsequently with an optical microscope. The size obtained from the SEM image correspond better to the size obtained from the HOLIMO image than the size obtained from the optical microscope image. The first two sizes agree within some % whereas the size from the optical image differs about 10% from the SEM and The HOLIMO value. The first and second row on figure 5.3 (b) show pictures from the same analogues taken with HOLIMO I. Their determined sizes agree within some % except the sizes from the images of a big hexagonaly structured crystal disagree almost 10%.

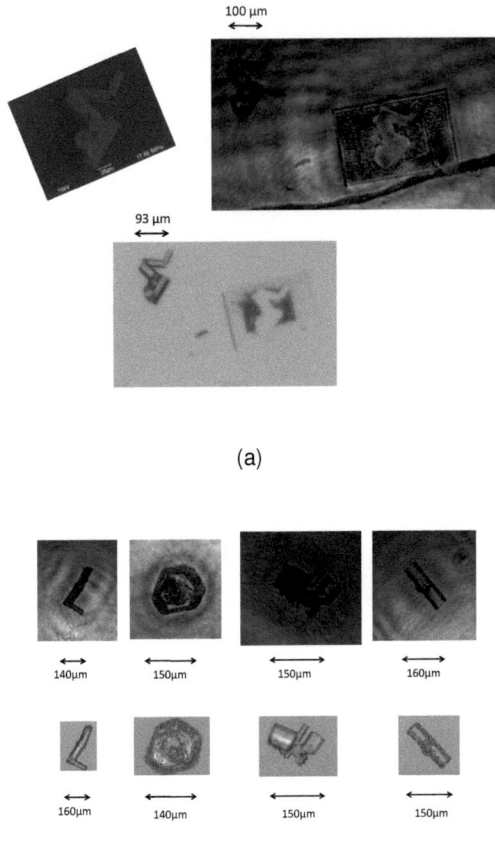

Figure 5.3: *Ice analogues seen with SEM and an optical microscope vs. holographic imaging. Panel (a) shows a SEM, a holographic reconstructed and an optical microscope image clockwise. The first row on panel (b) shows four reconstructions equivalent to the second row that shows images of an optical microscope from the same objects.*

5.4 Tests with HOLIMO II

For HOLIMO II, the same procedure like for HOLIMO I needed to be applied. Therefore, PSL spheres of a size 20.2 ± 0.4 μm with a standard deviation of 0.76 μm were used to obtain the right distances l and magnification M (figure 5.4). ± 0.4 μm is the uncertainty of measurement of the manufacturer. Figure 5.4 shows that not all of the spheres are equally large. In fact, the bigger ones indicated with the red line show 27% bigger sizes. 61 PSL particles have been investigated from different reconstructions and 11 of them showed such big sizes. This means that about 18% of the investigated particles differ in size from the standard sizes which was measured to be 22.0 ± 1.4 μm. The bigger PSL spheres were measured to have a size of 28.0 ± 1.0 μm. The smaller sizes were attributed to the standard size that should be 20.2 μm with a standard deviation of 2%. The averaged smaller size differ about 10% from the averaged manufacturers value. The varieties of the values found in the reconstruction show larger discrepancies for reconstruction planes closer to the camera or the PS. They showed a slight under- or overestimation of the sizes respectively. The overestimation was dominant due to the conical V_{obs}. That means that more particles were observed closer to the camera. Additional reconstruction planes of PSL spheres at different positions from the camera were investigated exemplarely in order to obtain an overall correction factor of 1.15 for the under- andoverestimation of sizes. This additional study was performed since most of the previously mentioned 61 PSL spheres were found inside the same reconstruction plane.

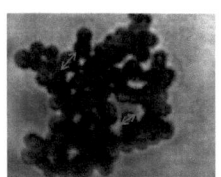 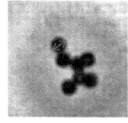

Figure 5.4: *This figure shows reconstructed images from 20.2μm PSL spheres used for calibration. The spheres do not show a uniform distribution overall. Most of the observed PSL spheres were of the same size like the ones in the clusters on the right and left. Those 'regular' PSL (blue arrow) were sometimes accompanied by bigger ones (red arrow). The object indicated with the red arrow measures 27% more than the one indicated with the blue arrow.*

5.5 ZINC measurements with HOLIMO II

In the end of the validating phase for the HOLIMO instruments, a measurement of a sample flow from ZINC was performed ([215]) with the improved setup of HOLIMO II. ZINC is a continuous flow diffusion chamber. This chamber is build out of two parallel plates that are covered with an ice layer in order to obtain supersaturation with respect to ice inside the chamber via the temperature difference between those two parallel walls. An aerosol flow is confined in the middle of the two walls via a sheath air flow such that the maximum of the ice supersaturation is obtained where the aerosol flow passes through the chamber. Consequently ice nucleation appears and the activated fraction (ratio of the amount of nucleated ice and the amount of measured particles at the inlet) is determined. All particles measured that reveal sizes bigger than 2 μm are supposed to be ice crystals and they show activation for activated fractions above 1%. ATD (Arizona Test Dust) particles with a concentration around 1000 cm^{-1} were produced with a fluidized bed aerosol generator and size selected at 800 nm with a SMPS. Those aerosols were directed towards a flow splitter and introduced equally into a CPC and ZINC. An OPC was applied after ZINC in order to compare the activated fractions from the CPC and the OPC. Measurements were performed at different T and RH_{ice}.

Figure 5.5 shows that HOLIMO II is capable of measuring sample flows as a stand alone instrument. Hydrometeor habits are categorized with respect to the temperature and supersaturation with respect to ice over a time period of about 70 min for this test experiment. The hydrometeors observed and shown here were as small as 7 μm at relatively low supersaturations $\leq$1.8%. The activated fractions at the time were the images have been taken were probably lower than 1% because [216] found activation fractions of several percents for ATD at the indicated T of -32 and -47°C. Hence we can say that there was no significant activation.

Overall the instrument works reliably. Problems though occurred when considering the speed of the reconstruction routine and the overhead produced due to constantly recording V_{obs} with a time resolution of 4 frames per second. First, 4 frames per second is too slow to give quantitative but it is fine for qualitative statements from the results. Second, due to the size of the original holograms, the amount of recorded data is immense. It would be meaningful to use a triggering device. This could be done by using an additional laser as a light barrier. An other possibility would be the use of a software trigger. A background image should be defined at wishes and recording of a subsequent image would only take place when the overall mean brightness or a certain

5.5. ZINC measurements with HOLIMO II

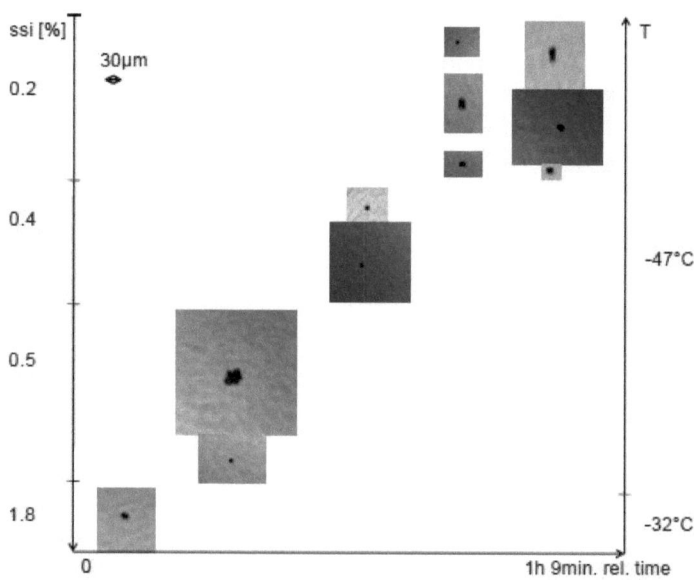

Figure 5.5: *Example pictures of a ZINC experiment classified with respect to temperature, ice supersaturation and experiment time. Interestingly, there seems to be an ice crystal with rosette like habits of about 30 μm.*

amount of pixels is change compared to the background image. For certain, the speed of the reconstruction routine needs to be improved. It is unclear whether this could be achieved in Matlab or whether a different application must be considered.

Chapter 6

Experimental and modelling results from AIDA measurements

After testing HOLIMO thoroughly in the lab and after HOLIMO proved to work reliably under laboratory conditions it was taken to three different campaigns in December 2007 and 2008 and in March 2009. The first two campaigns (IN11 and HALO02) took place at the Institute of Meteorology and Climate Research at the Forschungszentrum Karlsruhe in Germany where HOLIMO was applied to the AIDA chamber. The objectives were to compare optical instruments and obtain insight into the relation of pristine ice crystals and their ability to depolarize light.

The following two subchapters were adapted for a paper publication by Amsler et al. (2009) ([237]).

6.1 AIDA facility

The HOLIMO detector described in section 4 has been applied in the ice nucleation campaign IN11 which has been conducted in November 2007 at the cloud simulation chamber AIDA ([217]) of Forschungszentrum Karlsruhe.

These experiments aim to improve our knowledge of the ice nucleation and growth processes in atmospheric mixed-phase and ice clouds. Such clouds can be simulated in AIDA by controlled expansion cooling experiments which mimic the adiabatic expansion cooling of rising air parcels in the atmosphere (figure 6.1).

Experiments can be performed over a broad temperature range down to 183 K. The huge chamber volume of 83 m^3 thereby enables cloud lifetimes of up to 30 minutes. During IN11 several specific ice cloud characterization experiments have been performed in addition to the ice nucleation studies. One of these experiments (IN11_2) is

68   Chapter 6. Experimental and modelling results from AIDA measurements

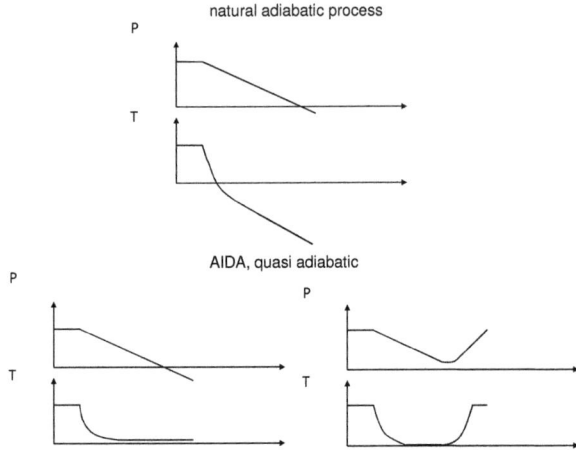

Figure 6.1: *Adiabatic expansion in the AIDA chamber. The sketch on top shows natural adiabatic process behavior. As long as the pressure is decreasing the temperature will decrease also. The two sketches at the bottom of this figure illustrate such processes and how they occur inside the AIDA chamber. The temperature levels off after a certain amount of time although the pressure is still decreasing. This is because of the influence of the walls. If the pressure is increasing then also the temperature increases.*

discussed in the present paper in terms of the dependence of the linear depolarization ratio on the microphysical details of plate-type ice crystals. Before the experimental results are discussed in section 6.2, the following two subsections describe briefly two basic AIDA instruments that are employed routinely in those ice cloud experiments, namely the optical particle counter WELAS and the laser light scattering and depolarization instrument SIMONE.

6.1.1 WELAS

A commercial optical particle counter WELAS (WhitE Light Aerosol Spectrometer, [217]) is mounted below the chamber inside the thermostated housing (see figure 6.2). It sam-

6.1. AIDA facility

ples chamber air through a vertical tube at a flow rate of 5 $l \cdot min^{-1}$. The OPC is used to measure number density and size distribution of liquid particles in the size range from 0.5 to 48 μm. The detection volume of WELAS is illuminated by a white light Xenon lamp. A photomultiplier detects the light which is scattered into angles from 78 to 102°. Particle number density can be obtained from single particle count rate and sampling flow rate through the optical detection volume. On the basis of Mie-theory calibration curves were deduced which deliver the relationship between scattered light intensity and particle diameter for spherical particles with a given refractive index. The use of a white light source ensures an unambiguous correlation between scattered light intensity and particle diameter. In contrast to spherical particles, WELAS cannot size classify ice crystals because the apparent size of an ice crystal depends on the accidental orientation of the crystals in the detection volume. However, WELAS can well distinguish between larger ice crystals and droplets staying at smaller sizes (i.e. WELAS can show the onset of ice nucleation).

Chapter 6. Experimental and modelling results from AIDA measurements

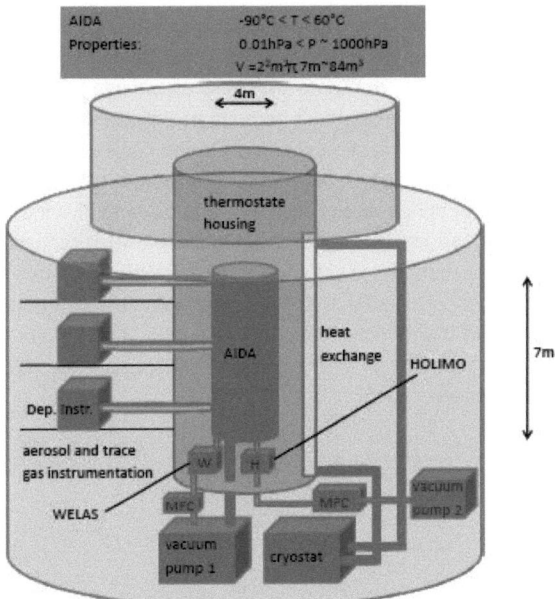

Figure 6.2: *Sketch of the AIDA facility. AIDA itself is the inner most cylinder. It has a diameter of 4 m and is 7 m high. This corresponds to a volume of approximately 84 m^3. It is surrounded by a thermal housing and aerosol and trace gas instruments. The wall temperature of AIDA is adjusted via heat exchange controlled by a cryostat in the basement. The temperature can be set between $-90°$ C and $+60°$ C. The inner temperature is controlled via adiabatic expansion with a vacuum system in the basement. The pressure can be set between 0.01 hPa and 1000 hPa. The vacuum system is also used for various sampling streams which are drawn off from the bottom of AIDA and controlled with a MFC. The point where HOLIMO was inserted into the measuring flow is indicated.*

6.1.2 Depolarization Instrument

The in situ laser light scattering and depolarization instrument SIMONE (Scattering Intensity Measurements for the Optical Detection of Ice) is described in detail elsewhere ([218]). Here, we give only a rough description of the set up and the measured quan-

6.1. AIDA facility

tities. SIMONE uses a cw semiconductor laser (λ= 488 nm, 10 mW) to generate a polarized and collimated light beam which is directed horizontally along the diameter of the AIDA chamber. The linear polarization state of the incident light beam can be adjusted by using a liquid crystal polarization rotator in front of the laser head. It is usually aligned either parallel or perpendicular to the scattering plane which is defined by the light beam and the overlapping detection apertures of two telescopes. The telescopes probe scattered light from the chamber interior at 2° in forward direction and at 178° in backward direction. The intersection between the laser beam and the telescope apertures in the center of the chamber defines the detection volume of the instrument of about 7 cm³. While the intensity of forward scattered light is measured directly by a photomultiplier, the polarization state of the backscattered light is analyzed by using a Glan-Taylor prism prior to the detection by two photomultipliers. In this way the parallel ($I_{\|}$) and perpendicular ($I_{\perp}$) intensity components with respect to the scattering plane are measured. From these experimental quantities, the averaged linear depolarization ratios $\delta_{\|}$ and $\delta_{\perp}$ of the nucleated ice crystals for incident light polarized parallel or perpendicular to the scattering plane are deduced:

$$\delta_{\|} = \frac{I_{\perp} - I_{\perp}^{bs}}{I_{\|} - I_{\|}^{bs}} \quad \text{for parallel polarized incident light}$$

$$\delta_{\perp} = \frac{I_{\|} - I_{\|}^{bs}}{I_{\perp} - I_{\perp}^{bs}} \quad \text{for perpendicular polarized incident light.}$$
(6.1)

$I_{\|}^{bs}$ and $I_{\perp}^{bs}$ are the backscattered intensities polarized parallel and perpendicular to the scattering plane. In this way, the linear depolarization ratio can be determined with an accuracy of 0.05 at a temporal resolution of 1 s. To compare with modelling results we need to express the depolarization defined in equation (6.1) by elements of the scattering matrix. By means of the definition of the Stokes vector ([219]), the intensities in equation (6.1) can be written as follows,

$$I_{\|} - I_{\|}^{bs} = \frac{I_{sca} + Q_{sca}}{2}$$

$$I_{\perp} - I_{\perp}^{bs} = \frac{I_{sca} - Q_{sca}}{2},$$
(6.2)

where I_{sca} and Q_{sca} are the elements of the Stokes vector of the scattered light. Next, I_{sca} and Q_{sca} can be obtained by applying the scattering matrix to the incident Stokes vector:

$$I_{sca,\parallel} = \frac{S_{11} + S_{12}}{R^2}; \quad I_{sca,\perp} = \frac{S_{11} - S_{12}}{R^2}$$
$$Q_{sca,\parallel} = \frac{S_{12} + S_{22}}{R^2}; \quad Q_{sca,\perp} = \frac{S_{12} - S_{22}}{R^2}.$$
(6.3)

The constant R is the distance from the scatterer. Use has been made of the relation $S_{21} = S_{12}$, which is valid for random orientation. Inserting eqs. (6.2) and (6.3) into equation (6.1) gives

$$\delta_\parallel = \frac{S_{11} - S_{22}}{S_{11} + 2S_{12} + S_{22}} = \frac{1 - S_{22}/S_{11}}{1 + 2S_{12}/S_{11} + S_{22}/S_{11}}$$
$$\delta_\perp = \frac{S_{11} - S_{12}}{S_{11} - 2S_{12} + S_{22}} = \frac{1 - S_{22}/S_{11}}{1 - 2S_{12}/S_{11} + S_{22}/S_{11}}.$$

6.2 AIDA Experiment IN11_2 from November 2007

As mentioned in the previous section the results of experiment IN11_2 of the AIDA 2007 campaign are discussed here exemplarily. The specific experiment discussed here was started at an initial gas temperature of 252.5 K and at nearly ice saturated conditions (both quantities still levelled off from the previous expansion). At 2916 s experiment time water droplets were injected into the AIDA volume by an atomizing nozzle. The water droplet injection lasted for 148 seconds indicated by the thin vertical dotted lines in figure 6.3. Dry ammonium sulphate background aerosol with a number concentration of about 2500 cm^{-3} has been pre-added to the chamber. A fraction of the supercooled water droplets starts to nucleate ice during the injection probably by expansion cooling and/or collisions with the background aerosol resulting in a localized mixed-phase cloud. During its spreading into the AIDA volume the mixed-phase cloud rapidly converts to a pure ice cloud by the Bergeron-Findeisen process. HOLIMO was used in IN11 to characterize the habit distribution of the ice particles in the simulated clouds. Figure 6.2 shows the setup of the AIDA chamber and the position of measurement of HOLIMO.

6.2. AIDA Experiment IN11_2 from November 2007

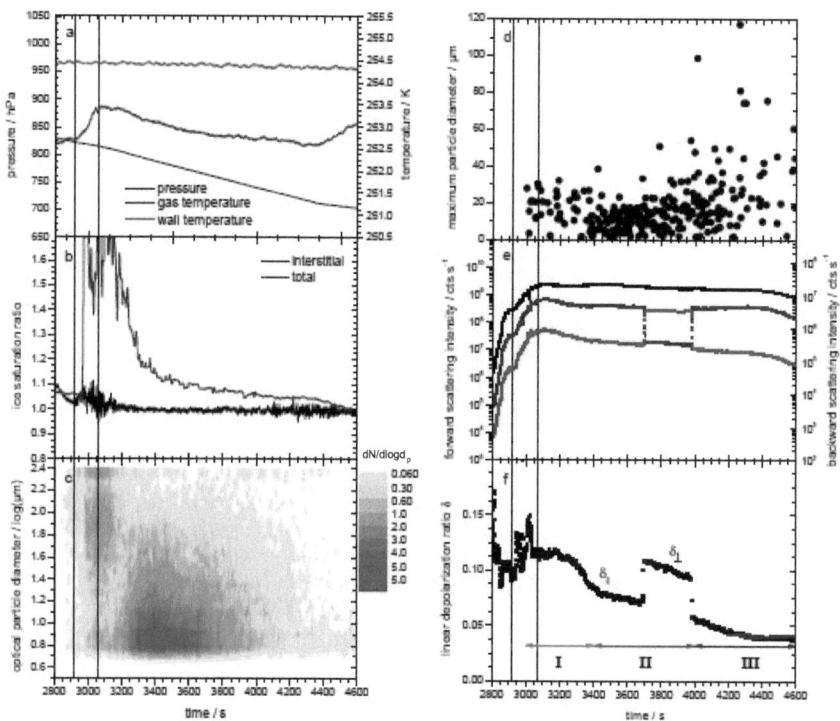

Figure 6.3: *Combined results from AIDA mixed phase cloud experiment 2 of IN11. Panel a shows the temperature of the wall of AIDA and the gas it contains and also its pressure. Panel b shows the ice saturation ratio of the total and interstitial water content inside AIDA. Panel c (color bar in dN/dlogd$_p$) and d show the WELAS and HOLIMO size distribution respectively. Panel e shows the backward scattered signal of the perpendicular and the parallel channel in blue and red with respect to the total forward scattered signal in black. Panel f shows parts of the linear depolarization ratio of the 2 backward channels. The vertical lines throughout all panels indicate the time of droplet injection.*

74　Chapter 6. Experimental and modelling results from AIDA measurements

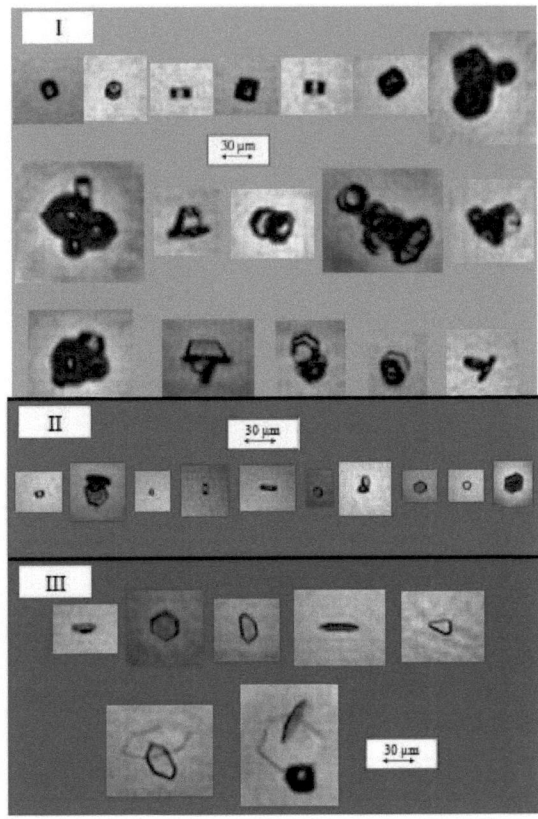

Figure 6.4: *Ice crystal habits of experiment 2 of IN11 during three different time slots showing three different phases of habits and frequency of occurrences.*

It is placed at the bottom of the chamber with a strictly vertical sampling line in order to avoid particle loss due to sedimentation. Figure 6.3 depicts p, T, s_i, WELAS and HOLIMO size distributions, scattering intensities and depolarization ratio data on the panels a, b, c, d, e and f respectively. Temperature and s_i values are given with an accuracy of ± 0.3 K and ± 5 % accordingly. The sizes HOLIMO can see are given with an error of $\pm 15\%$ (see subsection 4.4 theory of operation).

Panel a shows the behavior of the mean gas temperature inside AIDA during the expan-

6.2. AIDA Experiment IN11_2 from November 2007

sion cooling experiment. While the wall temperature of the vessel stays rather constant throughout the whole experiment, the droplet injection (region between vertical lines) leads to an increase of the mean gas temperature. After this event one can observe a constant decrease of the gas temperature until the point where the expansion cooling levels off due to the heat flux from the warmer chamber walls.

Panel b shows the ice saturation ratio with respect to the interstitial and total (interstitial and particulate) water contents inside AIDA. The injection of supercooled water droplets is clearly visible in an increase of the particulate water content. Meanwhile, the emerging ice cloud reduces the interstitial water vapor content and confines the saturation ratio to ice saturated conditions after a short period of enhanced noise during droplet injection. The interstitial phase remains close to a supersaturation ratio of 1 after the injection.

Table 6.1: Frequency of occurrence of ice crystal habits in 3 different periods of experiment IN11_2 (see figures 6.3 and 6.4).

period:	I	II	III
identified particles [%]	81	36	78
thin plates [%]	5	68	93
thick plates [%]	74	28	0
aggregates of thin plates [%]	0	2	7
aggregates of thick plates [%]	22	2	0

Panel c is obtained from the WELAS optical particle counter and reveals 2 modes of particle growth. The first and bigger mode with respect to the optical particle diameter emerges during the time of injection. The second mode grows directly after the vanishing of the first mode. It is very likely that this bi-modal size distribution reflects the spray characteristics of the atomizing nozzle. As already mentioned, the droplets were probably freezing due to the expansion of the spray. Some of the ice crystals may have grown fast to thick plates at relatively high supersaturation with respect to ice and close to water saturation within undiluted sections of the cloud of the spray. They may have fallen out when others were transported into regions with less or no supercooled droplets. Those crystals may grew slowly, due to a small but existing supersaturation with respect to ice, from very small crystals in the beginning to very large thin plates that aggregated at the very end of the experiment. An overview of ice crystal habits taken with HOLIMO supports this interpretation of the evolution of the measured size distribution (figure 6.4). In this thesis I am looking chiefly at the microphysical and optical properties of the existing ice cloud.

The maximum particle diameter deduced from the HOLIMO images (panel d) shows bigger particles during period I from 3000 to 3400s than during period II from 3400 to 4000s. Nevertheless, particles detected in period II are smaller than the ones observed during period III from 4000 to 4600s. 81% of the particles found within period I were identified. They split into 5% thin, 74% thick plates and 22% aggregates of thick plates. Most of the particles emerging from period II were either too small for HOLIMO to resolve their habits or they were out of focus. Therefore, the automated routine only identified 36% particles. Among these particles 68% thin and 28% thick plates have been found. In period III, 78% of the observed particles were identified. 93% of them were thin plates and 7% aggregates of them. Table 6.1 summarizes the findings in rounded figures.

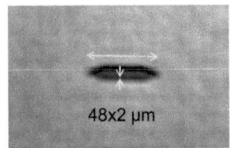
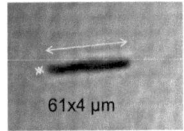
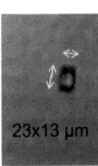

Figure 6.5: *Example pictures of aspect ratios* $\chi = maximumlength/thickness$ *for thin and thick plates. The example on the left hand side of a thin plate seen under grazing incident has a* χ *of 24. The thin plate on the right hand side has an apparent* χ *of 15 though the contribution of edge blurring is somewhat bigger than for the previous example. This is due to the fact that the effect of both the forefront and its opposite add up. The thick plate has a comparable low* χ *of 2.*

This habit distribution leads to an interesting result substantiated in panels e and f. They show the scattered intensity (total, forward and backward) and the linear depolarization ratios $\delta_\|$ and $\delta_\perp$. The depolarization ratio for parallel incident laser polarization $\delta_\|$

6.2. AIDA Experiment IN11_2 from November 2007

reveals very low values between 0.11 for the thick plates shortly after the droplet injection and 0.04 for the thin extended plates towards the end of the experiment. Between 3700s and 4000s the incident laser polarization was changed to be directed perpendicular to the scattering plane. The corresponding depolarization ratio $\delta_\perp$ is offset by about 0.04 with respect to $\delta_\parallel$. Even more striking than these low values is the trend of both quantities. $\delta_\parallel$ starts off in the beginning of period I around 0.1 indicating first thick plates and aggregates conglomerated of them. The depolarization ratio then decreases after thick plates and aggregates start to disappear (first mode of the WELAS data). This trend is pursued because the fraction of thick plates becomes smaller and the fraction of thin plates bigger. Additionally, during period III, the thin plates were getting bigger and eventually aggregate conglomerated out of them. To understand this behavior by comparison with the results from ray tracing calculations described below in this section, it is important to know the ratio χ of maximum length versus thickness of the plates. Pictures of thin plates under a grazing angle and thick plates were analyzed in order to determine χ. Figure 6.5 shows some χ values assigned to thick and thin plates on a few examples pictures. It is better to use thin plates under a grazing angle because the effect of edge blurring of both the forefront and its opposite add up if only one of the two opposite thin prism facets of the plate is seen. It was found that χ for thick plates ranged from 1.1 to 4.1 with a standard deviation of 0.9 and for thin plates from 9 to 65 with a standard deviation of 14.7.

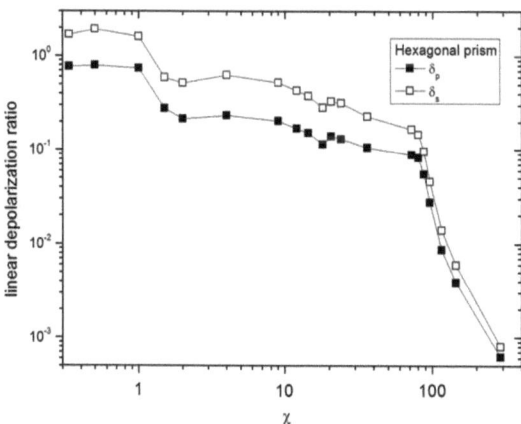

Figure 6.6: *Linear depolarization ratios vs.* χ.

Chapter 6. Experimental and modelling results from AIDA measurements

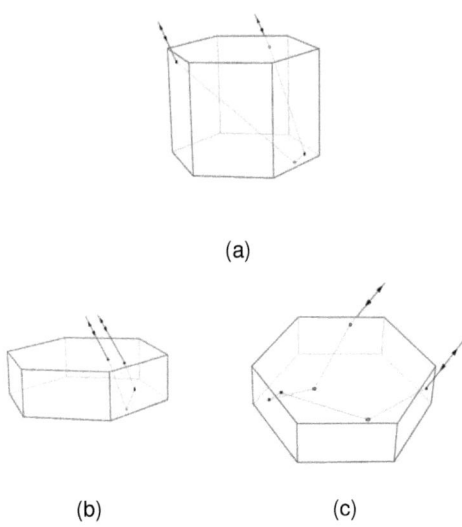

Figure 6.7: *Ray paths which contribute most to linear depolarization for* $\chi < 1.5$ *(a)*, $2 < \chi < 18$ *(b) and* $\chi > 18$ *(c)*.

In order to assist the interpretation of the depolarization measurements, calculations were carried out using a geometric optics ray tracing program ([220]). In ice crystal prisms, internally reflected and refracted ray paths will be chiefly responsible for backscattering. These processes result in the reorientation of the incident polarization vector at every interface, leading to depolarization when the backscattered ray is transposed into the initial plane of polarization ([221]). Figure 6.6 shows the calculated δ_p and δ_s at 178° scattering angle for a range $1/3 < \chi < 288$. In agreement with the measurements, δ_p and δ_s decrease with increasing χ, and δ_s is larger than δ_p, although the modelled depolarization values are higher by about 0.1. This discrepancy might partly be due to the low thickness of the investigated ice crystals, which bring geometric optics to the margin of its applicability. Notable in figure 6.6 are three plateaus for $\chi < 1.5$, $2 < \chi < 18$ and $18 < \chi < 80$. These plateaus are caused by contributions from specific ray paths. The large depolarization for $\chi < 1.5$ is mainly due to the ray path shown in figure 6.7a: rays enter through a prism facet, are totally reflected at a basal facet followed by a reflection at the next but one prism facet from the entrance facet and refraction through the next but one facet. This ray path does not occur in prisms with larger χ. For $1.5 < \chi < 100$ the largest contribution to the intensity of the light scattered at 178° is due to rays entering and leaving through the same basal facet after being internally reflected at the opposite basal facet and one prism facet (figure 6.7b). These rays cause most of the linear depolarization for $\chi > 18$. However, in the region $2 < \chi < 18$ the largest contribution to depolarization results from rays entering through a prism facet and leaving the crystal at the next but one facet after four internal reflections, of which three are total internal reflections at the basal facets (figure 6.7c). For $\chi > 100$ depolarization decreases sharply due to the increasing number of rays which are reflected and refracted at basal facets only.

6.3 More experiments from the IN11 campaign from November 2007

In this section four of the experiments from the IN11 campaign November/December 2007 at the AIDA chamber at the Institute for Meteorology and Climate Science at the Research Center Karlsruhe are presented. IN11_1, _2 and _64 are experiments about heterogenous and IN_9 about homogeneous freezing processes in warm and mixed phase clouds. Those experiments are called by the name of the experiments performed at AIDA in order to relate best the data. In the following they are referred to as experiment 1, 2, 3 and 4. Especially experiment 2 was already covered in great detail in section 6.2. Here, it is put in context with the other 3 experiments mentioned above. 38%, 28%, 46% and 33% of the holograms of those experiment could be used for segmentation respectively.

6.3.1 Heterogeneous cloud experiments from IN11

Figure 6.8 of experiment 1 shows two regions of super cooled droplet injection starting off around -19° C. The particle concentration and RH_{ice} show an immediate increase as well as the HOLIMO size distribution and habit diagram. It indicates clearly a phase of droplet injection. Before experiment time 1500s all values drop to 0. Around 1700s after the start of the experiment a second short period of droplet injection took place. In the following, lots of big thin hexagonal plates appeared in the size range of 2 to 140 μm.

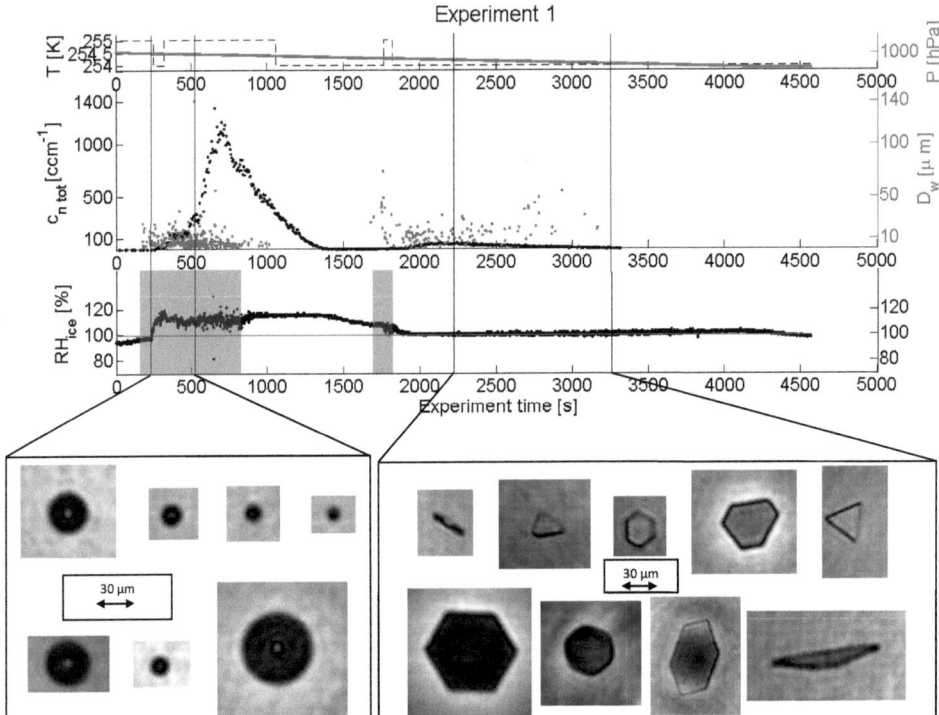

Figure 6.8: *Time evolution of ice crystal habit, relative humidity, pressure and temperature profile of the MPC experiment 1 in AIDA at 255 K. The upper most plot shows the temperature (dashed line) and pressure evolution (gray line). The next lower plot shows the total number concentration $c_{n\,tot}$ in cm^{-3} (black points) recorded with WELAS for a size range from 5 μm to 200 μm. This instrument is an optical particle counter that is calibrated for spherical particles. This plot also shows the size proxy D_w in μm (gray points) from HOLIMO. It shows the time span of observable hydrometeors with WELAS and HOLIMO respectively. The smallest hydrometeor found with HOLIMO was 2 μm and the biggest 140μm. The gray shaded regions ($187 \leq t_{exp} \leq 806$s and $1682 \leq t_{exp} \leq 1757$ s) in the RH_{ice} profile indicate supercooled water droplet injection. Representative samples of HOLIMO images of different classes are shown underneath the RH profile. Predominantly droplike and regular hydrometeors were recorded with HOLIMO and they are shown independently with respect to time from the top left to the bottom right corner.*

6.3. More experiments from the IN11 campaign from November 2007

Figure 6.9 of experiment 3 shows one region of super cooled droplet injection starting off around -20° C. RH_{ice} shows a fast increase due to fast adiabatic expansion of the volume of the AIDA chamber. Both WELAS and HOLIMO show a bimodal size distribution. The habit diagram of HOLIMO reveals smaller objects in the size range of 2 to 50 μm. The habits found were mostly small thick plates, lots of very small spherical like objects and some irregular maybe splintered ice crystals.

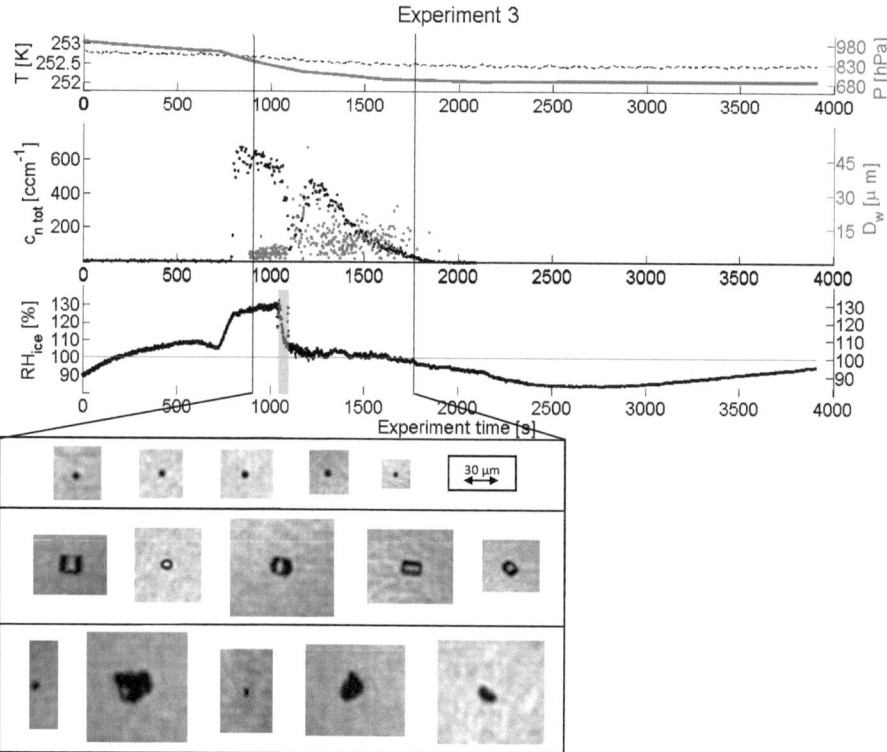

Figure 6.9: *As figure 6.8, but for experiment 3 with sizes between 2 μm and 50 μm. Supercooled water droplet aerosols were injected during the experiment indicated by the green shaded region. Droplike, regular and irregular hydrometeors were recorded with hOLIMO and they are shown independently with respect to time from the left to the right corner.*

Chapter 6. Experimental and modelling results from AIDA measurements

Table 6.2: Frequency of occurrence of hydrometeor habits for three different size ranges for experiment 1. Size range I: 0-38 μm. Size range II: 38-76 μm. Size range III: 114-152 μm. The percentages are rounded figures. Therefore, it is possible that they add up to more or less than 100%.

T=255K	Droplikes [%]			Regulars [%]			Irregulars [%]			number of images		
RH [%]	I	II	III	I	II	III	I	II	III	I	II	III
82-100	1	0	0	9	0	0	6	17	0	66	2	0
100-110	3	0	0	31	34	0	8	50	100	175	10	1
110-120	2	0	0	38	0	0	1	0	0	169	0	0
120-130	0	0	0	1	0	0	0	0	0	4	0	0
82-130	6	0	0	80	34	0	15	67	100	414	12	1

Table 6.3: Frequency of occurrence of hydrometeor habits for three different size ranges for experiment 2. Size range I: 0-75 μm. Size range II: 75-150 μm. Size range III: 150-225 μm. The percentages are rounded figures. Therefore, it is possible that they add up to more or less than 100%.

T=255K	Droplikes [%]			Regulars [%]			Irregulars [%]			number of images		
RH [%]	I	II	III	I	II	III	I	II	III	I	II	III
95-100	5	0	0	42	0	0	32	33	0	262	1	0
100-110	1	0	0	12	0	0	7	67	100	68	2	1
95-110	6	0	0	56	0	0	39	100	100	330	3	1

Tables 6.2, 6.3 and 6.4 show the frequencies of occurrence of experiment 1, 2 and 3.

6.3. More experiments from the IN11 campaign from November 2007

Table 6.4: Frequency of occurrence of hydrometeor habits for one size range for experiment 3. Size range I from 0-60 μm. The percentages are rounded figures. Therefore, it is possible that they add up to more or less than 100%.

T=253K RH [%]	Droplikes [%]	Regulars [%]	Irregulars [%]	number of images
85-100	0	5	2	30
100-110	7	41	10	218
110-120	0	2	0	6
120-130	0	31	1	120
85-130	7	79	13	374

6.3.2 Homogeneous cloud experiments from IN11

Figure 6.10 of experiment 4 starts off around -43° C. RH_{ice} shows a fast increase up to 60% due to fast adiabatic expansion of the volume of the AIDA chamber. Both WELAS and HOLIMO show a steep ascent in the size distribution after the maximum in RH_{ice} was reached. The habit diagram of HOLIMO reveals smaller objects in the size range of 2 to 43 μm. The habits found were very irregular.

84   Chapter 6. Experimental and modelling results from AIDA measurements

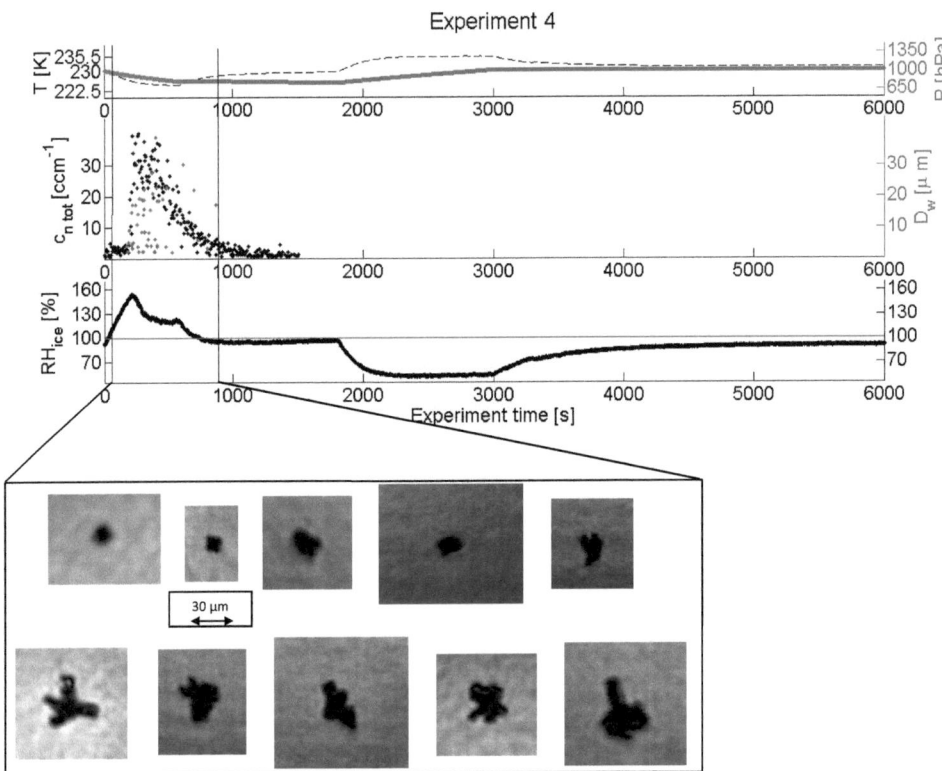

Figure 6.10: *As figure 6.8, but for experiment 4 with sizes between 2 μm and 43 μm. Only irregular hydrometeors were recorded with HOLIMO. They are shown with respect to time from the top left to the bottom right corner.*

6.3. More experiments from the IN11 campaign from November 2007

Table 6.5: Frequency of occurrence of ice crystal habits for one size range for experiment 4. Size range I from 0-43 μm. The percentages are rounded figures.

T=230K RH [%]	Regulars [%] I	Irregulars [%] I	Number of images I
51-100	4	0	2
100-110	4	0	2
110-120	2	0	1
120-130	52	12	32
130-140	4	2	3
140-150	20	0	10
51-150	86	14	50

Table 6.5 shows the frequency of occurrence of experiment 4. Figure 6.11 summarizes the findings of the frequency of occurrence for those 4 experiments in panel (a), (b), (c) and (d) respectively. Table 6.6 summarizes the parameters of the four different IN11 experiments.

Table 6.6: Description of the 4 different IN11 experiments.

	start temperature	seeds	RH_{ice} range
Exp1	254.5K	supercooled water and sulfuric acid	97% - 109%
Exp2	255K	supercooled water and ice	95% - 107%
Exp3	252.75K	supercooled water and sulfuric acid	85% - 132%
Exp4	230	sulfuric acid	50% - 155%

These findings of the three heterogeneous experiments agree with the habit diagram for in cloud crystal growth (figure 1.1 from chapter 1). Habit imagery of HOLIMO II for the temperature and RH range of experiment 1, 2 and 3 are within the plate regime of figure 1.1. Most of the times RH_{ice} is a lot smaller than 10% and therefore the dominating habits are thin plates. The duration of RH around 10% was longer for experiment 1 than for experiment 2 and even longer than for experiment 3. As a consequence, the habits of those experiments are similar but largest for experiment 1 and smallest for experiment 3. The average sizes of experiment 1, 2 and 3 (11, 17and 9 μm respectively) show a slight difference to the observed maximal size. Experiment 1 shows the biggest

Chapter 6. Experimental and modelling results from AIDA measurements

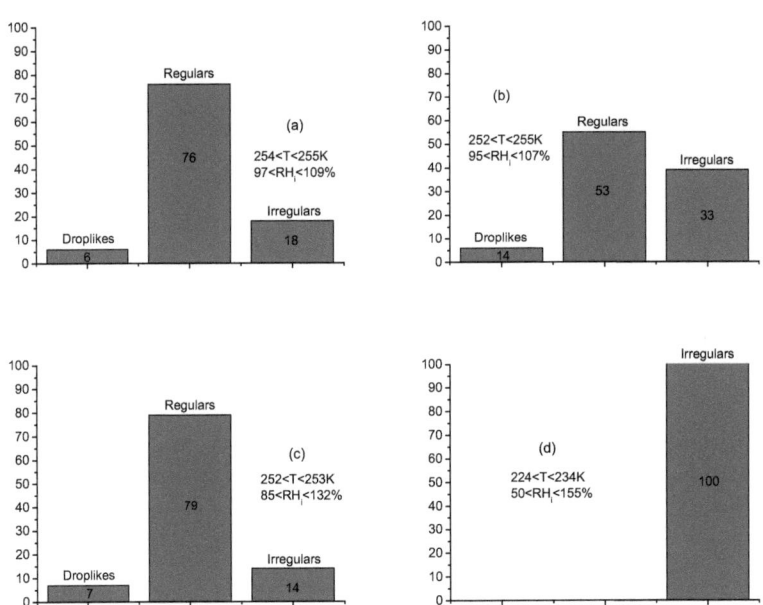

Figure 6.11: *Frequency of occurrence of hydrometeor habits for sizes <60 μm for the 4 different IN11 experiments.*

sizes but experiment 2 shows more large ice crystals due to enhanced aggregation of thin plates. Experiment 4 was a homogeneous experiment and out of range of the habit diagram of figure 1.1. Very high supersaturations were reached during this experiment. Hence, the observed habits were solely irregular with an average size of 12 μm.

Warm clouds will probably evolve into mixed phase clouds at -18°C very similarly regarding the amount of droplike, regular and irregular habits of the hydrometeors. This evolution may be independent of the values reached by RH_i as long as it is higher than 100% (see table 6.6 and figure 6.11). This is because the warm clouds of experiment 1 and 3 were both initiated with supercooled water plus sulfuric acid. RH_i was >100% for longer time periods for experiment 1 than for experiment 3 but the latter one exhibited larger values of RH_i. Differences in this evolution are observed regarding the liquid water content and the duration of RH_i >100% because the time of droplet injection was

6.3. More experiments from the IN11 campaign from November 2007

more than one order of magnitude longer for experiment 1 than for experiment 3. The size and shape of the ice crystals evolved towards thin plates with a larger maximal diameter for a higher liquid water content and for longer periods of RH_i >100% compared to the thick plates occurring at lower liquid water content and for shorter periods of RH_i >100%. Experiment 2 was initiated with supercooled droplets and ice crystal seeds at around -18°C. This cloud type will probably evolve to a mixed phase cloud with an increased amount of irregular habits compared to the mixed phase cloud evolution of experiment 1 and 3 due to the initially used ice crystal seeds. A period of thick plates is followed by a period of thin plates. Remarkably, the irregular ice crystals are aggregates made out of thick or thin plates within those periods respectively. The change of constituent type and/or RH_i and/or liquid water content of a cloud during its lifetime can therefore alone change the habits of ice crystals inside a mixed phase cloud and hence its radiative properties. Experiment 4 shows that homogeneous freezing processes probably lead to a cold cloud evolution with irregular habits solely. Their size will only depend on the time and amount of RH_i >100%.

HOLIMO has been successfully tested in a measurement campaign at the Institute for Meteorology and Climate Science at the research center Karlsruhe in Germany. This study concentrates on data sets about mixed phase clouds initiated from supercooled water droplets. Very low values of the linear depolarization ratio below 0.12 have been found during such a cloud evolution event. The decreasing trend of the parallel channel of the linear depolarization ratio from 0.1 to 0.04 is accompanied by the increasing trend of χ from 2 to 20. These low and experimentally found values of the linear depolarization ratio match with the deduced aspect ratio χ of the ice crystals obtained from their habits detected by HOLIMO. To the best of our knowledge such low linear depolarization ratio values for randomly oriented thin ice crystal plates have never been reported to date. The results are supported by geometric optics ray tracing calculations for thin plates. Plates with the described properties show very low linear depolarization ratio values that can not be distinguished from those of water droplets. This implies the possibility that small values of depolarization not only stem from liquid clouds embedded in cirrus clouds [5] but also from regions of randomly oriented thin plates inside of them.

All the measurements described in this chapter were performed under controlled situations in the laboratory and laboratory like environments. The PINC II campaign finally allowed for measurements inside clouds. In the next chapter results obtained from this last campaign are discussed.

Chapter 7

Experimental results from PINC II field measurements

The last campaign discussed in this chapter of the thesis is the PINC II campaign from February/March 2009 at the high altitude research station on the Jungfraujoch in the Bernese Alps in Switzerland.

7.1 Introduction

In-cloud measurements have taken place since decades at the high altitude research station on the Jungfraujoch, among which the GAW program was dedicated to the measurement of atmospheric aerosols and gases for over one decade ([222]). Cloud microphysical parameters like cloud particle size ranges were very intensively studied during the Cloud and Aerosol Characterization Experiments (CLACE) from 2004 to 2006. A long term study by [223] revealed that the station is inside clouds 37% of the time in winter and 20% in summer ([224]). CLACE 4, which took place in February/March 2005, sampled mixed-phase clouds ([225]). Mixed-phase cloud occur at temperatures between 0 and -35°C. Since our aim is to investigate mixed-phase clouds we went to the Jungfraujoch in February/March 2009 for the PINC II campaign.

[226] investigated the patchiness in mixed-phase clouds, i.e. to what extent the cloud consists of separate regions containing only ice crystals or only cloud droplets. Figure 7.1 from [226] shows two possibilities of phase mixtures inside a mixed-phase cloud. The spatial homogeneity or inhomogeneity in mixed-phase clouds is important for the radiation balance and precipitation formation. The Bergeron-Findeisen process takes place in a mixed- phase cloud because ice crystals can grow at the expense of wa-

ter droplets due to the difference in saturation vapor pressure over ice and water. If ice crystals and cloud droplets co-exist in a given cloud region, ice crystals will grow rapidly due to the Bergeron-Findeisen process and fall out of the cloud. Hence, the degree of phase mixing influences the precipitation formation from the cloud. Airborne studies revealed that mixed phase clouds show a patchiness down to a spatial resolution of 100 m ([226]), which corresponds to the minimal time resolution of their instrument. A theoretical consideration ([227]) showed that the Bergeron-Findeisen process is most efficient in mixed phase clouds that show a complete mixture of ice crystals and droplets ([226]). [228] suggest that such a patchiness does not originate from a local inhomogeneity in an aerosol distribution but from dynamical processes within the cloud.

The following results considering patchiness of clouds related to fluctuations in T and RH for the Jungfraujoch experiment from March 1 starting 13:33 local time are work in progress. 3 hourly-data retrieved from observation by eye at the high altitude research station on the Jungfraujoch describe that it was in cloud at 9, 12, and 15 local time. At 18 local time there were alto cumulus clouds reported with a cover of 7/8 at 945 m ASL. Meiringen (637 m ASL) reported on that day that there was one layer of strato cumulus with 2/8 cover and at 1829 m ASL and one layer of alto stratus with 7/8 cover and at 3048 m ASL around the Jungfraujoch region at noon. The second layer is exactly at the same hight as the Jungfraujoch. The wind was coming from SE for the indicated hours.

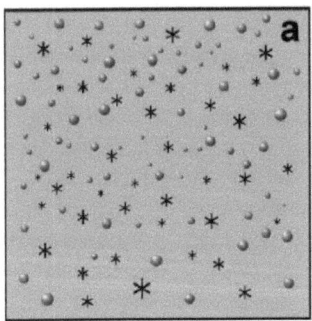 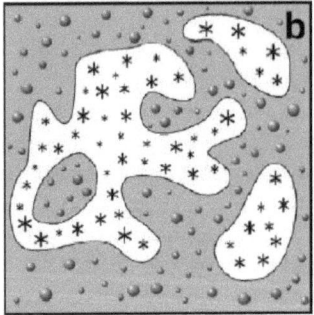

Figure 7.1: *Illustrations from [226] about the patchiness of clouds.* a *is a fully mixed phase cloud and* b *shows regions where there exist only droplets and others where there exist only ice crystals.*

7.2 Experiment and setup

Figure 7.2 shows the setup of HOLIMO on top of Sphinx of the high altitude research station on the Jungfraujoch. The instrument has a dimension of 600x400x240 mm. It is located to the left of the instrument in gold. The box is thermally insulated and contains the laser and its power supply and control unit. The camera and two self regulating heaters are also placed inside the aluminum box. It was tested for temperatures down to -70°. The measuring cell is attached to one side of the box. The inlet can be seen on the top of the aluminum box. The inlet of HOLIMO is a tube of a diameter of 10 mm and since the flow rate is 9.5 l min^{-1} this results in a flow speed of 4.03 m s^{-1} for laminar flow at the entrance of the inlet. At March 1 in the afternoon the average wind speed between 13:00 and 14:00 local time was 4.23 m s^{-1} which is almost the same as the sink velocity introduced for measurements with HOLIMO.

Figure 7.2: *This figure shows the HOLIMO (red arrow) instrument applied to the high altitude research station on the Jungfraujoch February and March 2009. It is placed next to the fog sampler in gold facing west.*

Several different data sources were included in the analysis of the Jungfraujoch campaign. Data series of temperature, relative humidity and windspeed were collected from the local meteorological service from the climap data base from MeteoSchweiz (figure 7.3), backward trajectories from the on-line HYSPLIT model (figure 7.4) related to temperature, relative humidity and wind speed ([229] and [230]) and CPC data from PSI were used to complement our analysis.

Climap data shows a wind speed of about 4.5 m s^{-1} from 13:30 until 13:45 local time during the experiment at relative humidities between 86 and 83% and temperatures between -8 and -9°C for Jungfraujoch. The HYSPLIT backward trajectory shows an air mass coming from the south, easterly from Marseille. It shows temperatures around -10°C and relative humidities between 95 and 98% along the trajectory at the experiment time.

7.2. Experiment and setup

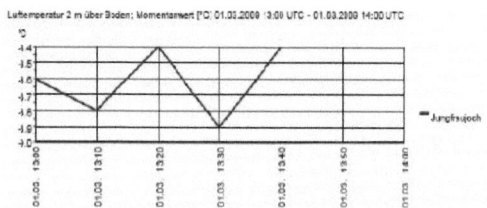

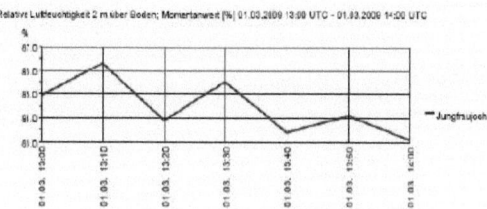

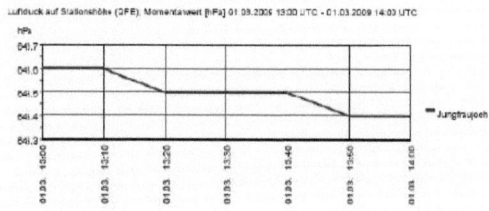

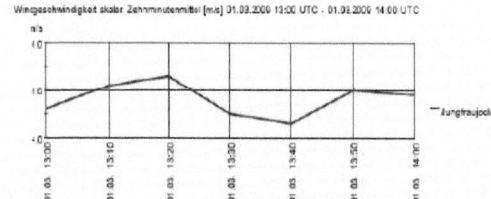

Figure 7.3: *Relative humidity, temperature, pressure and wind speed for March 1 2009 between 13:00 and 14:00 [courtesy of MeteoSchweiz].*

94   Chapter 7. Experimental results from PINC II field measurements

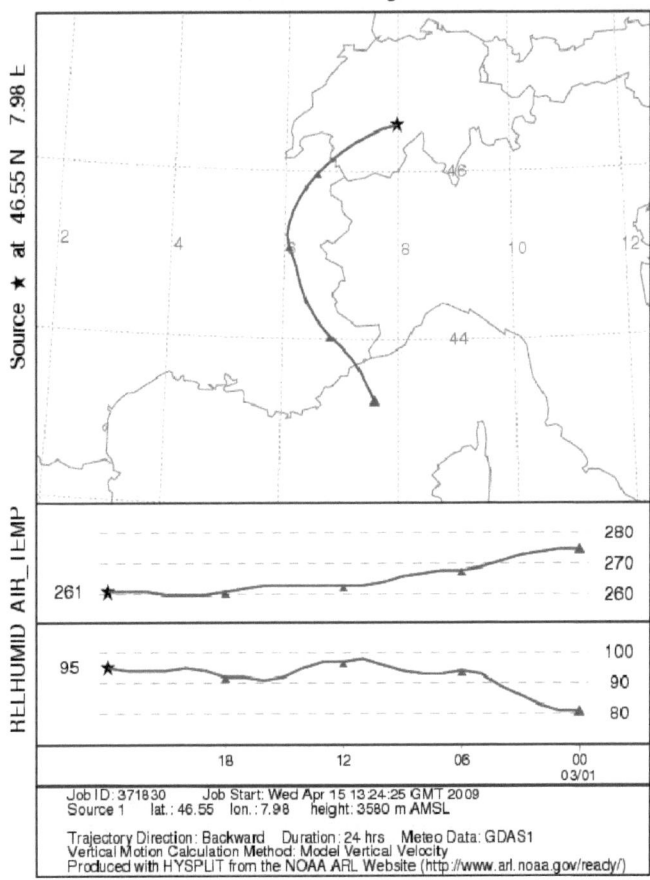

Figure 7.4: *Back trajectory calculated with the on-line HYSPLIT model with respect to ambient temperature and relative humidity ([229]) for March 1 2009.*

7.3 Preliminary results

The saturation ratio with respect to ice was below 1 throughout this experiment and therefore the ice crystals seen in figure 7.5 are sublimating. This is supported by the first, eighth and tenth picture of ice crystal habits from the top left corner on figure 7.5. They show rimed ice crystals made of pristine hexagonal plates that are sublimating from the edges. This behavior may also describe the evolution of the cloud.

The ice crystals seemed to have grown for a long time in a supersaturated region with respect to ice to sizes larger than 1 mm (not shown). None of the observed ice crystals show compactification or sintering which would be an indication of the crystals being blown from the ground towards the instrument. Additionally, they reveal pristine shapes

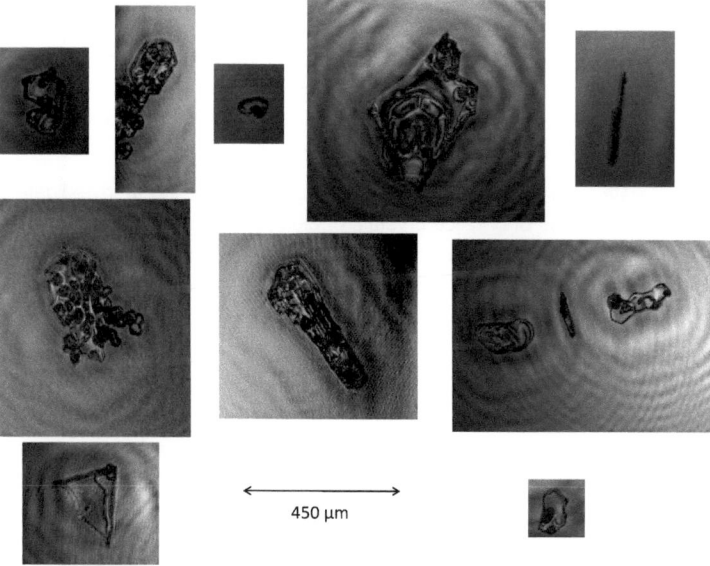

Figure 7.5: *Ice crystal habits from the PINC II campaign from March 1 2009 in the afternoon 13:33 - 13:46 local time. Pictures are shown in chronological sequence from top left to bottom right.*

which also indicate that they have grown and were measured inside the cloud. Their degree of riming suggests that they were in regions of updrafts or downdrafts where they collected water droplets. The ice crystals observed stem from different time periods. The time resolution of HOLIMO is 0.25 s. In this experiment from March 1 around 13:30 local time there were very distinct time intervals of droplets, ice crystals or nothing.

The transition between images with only ice crystals to only droplets often occurs at the minimal time resolution of 0.25 s. This would be a little more than 1 m spatial resolution. Sometimes multiple hydrometeors were found in a single frame. These multiple hydrometeors were always in the same phase either only ice crystals or only cloud droplets.

The interpretation of the data retrieved with HOLIMO is based on the standard interpretation of the interference pattern. The same principle is applied for instance in the SID3 (Small Ice Detector 3) sonde. Ice crystals reveal special or no symmetries in their interference pattern according to their habit. The interference pattern is very distinct. Droplets, on the contrary, reveal total symmetry within their interference pattern. This pattern shows the known Bessel intensity distribution for spherical apertures with relatively low contrast compared to the pattern exited by ice crystals. Therefore, ice crystals and droplets can be distinguished without performing a reconstruction.

Nevertheless, data from interference pattern can already be used to investigate if the cloud consists of patches of only ice crystals or cloud droplets or is well-mixed (figure 7.1). Problems can admittedly occur when the hydrometeors are very close to the resolution limit of HOLIMO and also when small ice crystals with spherical shape are oriented perpendicular towards the camera whilst being recorded.

The first results are obtained ignoring gaps between frames, because gaps can be caused either by a limited hitting rate of HOLIMO or could indicate cloud-free patches. All the possible images of consecutive droplets not interrupted by images of ice crystals are counted and vice versa for ice crystals. From this normalized histograms were produced and compared with modelled calculations for a complete mixture of hydrometeors inside a mixed phase cloud. HOLIMO detected ice crystals in 49% of the cases at the Jungfraujoch for the experiment described above. Therefore, a series of uniformly distributed random numbers was produced for the modelling of such a mixed phase cloud. Everything below 0.49 was attributed to the ice crystal and everything above to the droplet class (0 and 1 respectively). This sequence of 0 and 1 is treated the same way as the data retrieved by HOLIMO. A normalized histogram was produced and shown along the histograms retrieved from the HOLIMO data in figure 7.6.

7.3. Preliminary results

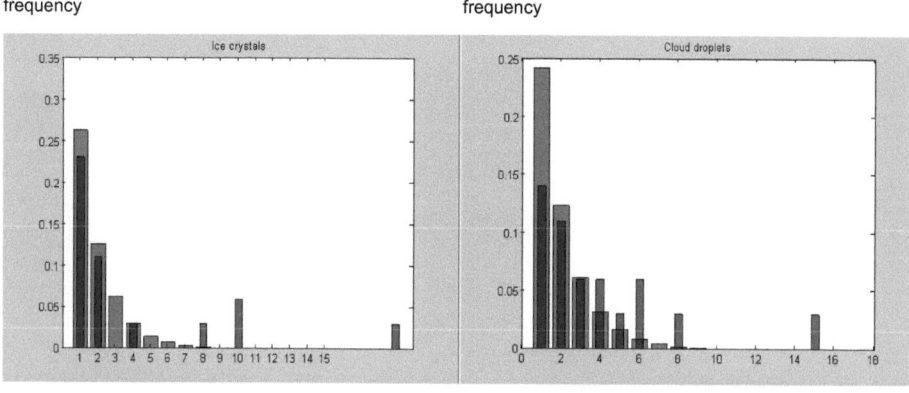

Figure 7.6: *Modelled random distribution of consecutive ice crystals and cloud droplets (blue bars) vs. measured distribution (red bars) on the left and right panel respectively. The probability to measure ice crystals was measured to be 0.5 and therefore set to $\leq$ 0.5 in the modelled random distribution.*

Most of the hydrometeors are seen in the lower bins indicating a complete mixture of the different hydrometeors. Nevertheless, HOLIMO sees larger contributions than the modelled data towards the larger bins indicating the presence of patches of the different hydrometeors.

In a next step we will investigate how to handle data gaps. We need to decide how to distinguish between gaps that are short enough and could be caused by the limited hitting rate of HOLIMO vs. those indicating cloud free areas. After that we should be able to decide whether our data indicates a fully-mixed cloud as in figure 7.1a or a cloud with distinct regions of either ice crystals or cloud droplets as in figure 7.1b.

Chapter 8

3 D modelling of ice crystal habits

The results obtained with a holographic instrument can also be treated for the purpose of 3D modelling of the observed objects. However, this technique is limited. There are several problems along the calculation.

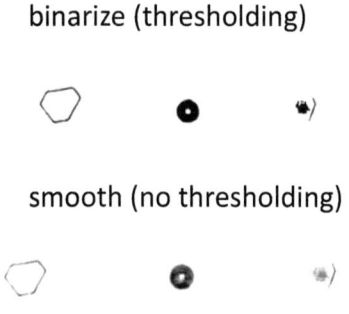

Figure 8.1: *Comparison between binarization with respect to a certain threshold and smoothing without. A thin deformed triangle, a droplet and a thin hexagonal were investigated. The object on the right hand side should be a thin hexagonal plate. In this case both binarization via thresholding and smoothing failed to deliver a true image of the object.*

Usually, there are lots of reconstruction planes collected throughout V_{obs} within the resolution limits of the system. Then the planes are binarized, stacked and made semi transparent. This enables to turn the object on the fly. Unfortunately, binarization depends on a threshold and can introduce noise. It also is only applicable for objects that are oriented with an angle of 0°. Intrinsic tilted features like holes would be assigned to the plane where they are in focus. Stacking can not be done if the object is big and tilted because the focal planes would have an offset among each other throughout the images. One possibility to avoid thresholding problems is to smooth the reconstructed image (figure 8.1). This means that a 4x4 window could be moved over the image in order to obtain a mean value for the first pixel in the upper left corner of the window. This can lead to very good images that resemble a binarized image without the ambiguity of thresholding. Figure 8.1 shows that the result is not all the times helpful. The object on the right hand side should be a thin hexagonal plate. Only parts of the object and noise in its center can be seen in both the binarized and the contrast image since the contrast in the vicinity of the plate is very low.

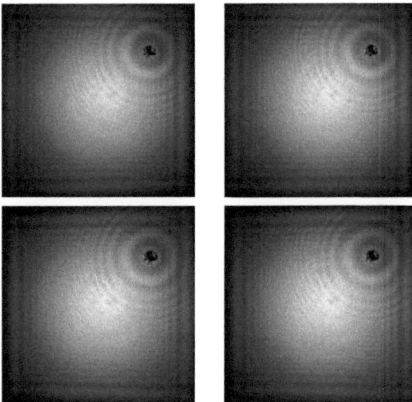

Figure 8.2: *3 D loop rotation around a reconstructed object obtained from a hologram around the inherent aperture angle of about 9°. One can see that the plane shown is tilted towards the right in the upper left corner and is moved to the left towards the lower right corner.*

In holography, there is also the possibility for the observer to move around the object. Those small changes in the point of perspective can be achieved by changing the parameter l in (E.1) along the image in the nominator of equation 3.2. This means that the reconstruction plane would be tilted with respect to the camera sensor plane. Nevertheless, in practice, the object can only be seen from different sides within the inherent angle of the holographic instrument i. e. the aperture angle. In the case of HOLIMO this angle is about 9°. Although being small, the change in the point of perspective can be witnessed in figure 8.2.

One other possibility would be to calculate iso caps and surfaces and interpolate between them. Here, again, a dependence of the binarization and the orientation of the object would lead to unsatisfying results. Figure 8.3 shows an example of the stacking method for a dendrite. The results are acceptable but look like a puzzle. The dendrite is seen under an angle of approximately 50°. Figure 8.3 also shows an example of the iso surface method for a droplet. It reveals that the reduced resolution in longitudinal direction distorts the droplet to a barrel. Additionally, figure 8.3 shows that the longitudinal resolution is 10 times worse than the lateral resolution. It can be corrected for by changing the aspect ratio of the figure but is still not satisfactory.

Those considerations show that it is possible to do a 3 D modelling of hydrometeors but every single objects needs to be modelled by hand due to the big variety of hydrometeor sizes, shapes and orientations.

Chapter 8. 3 D modelling of ice crystal habits

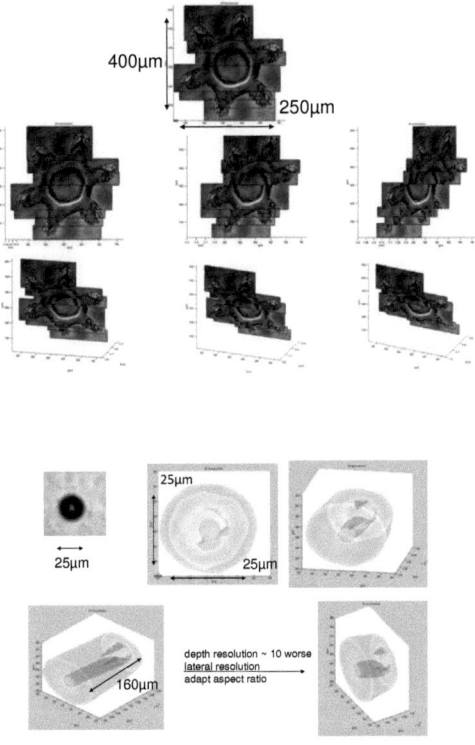

Figure 8.3: *3D modelling of a dendrite and a droplet. The dendrite is composed of all the parts in focus of the dendrite coming from different reconstruction planes in order to obtain a 3D model representation of the dendrite. The droplet is treated with isosurface and -caps in order to obtain a 3D model representation of the droplet.*

Chapter 9

Conclusion

Point Source (PS) digital in-line holographic microscopes are useful in-situ instruments in Atmospheric and Climate Science. It was shown that small hydrometeors down to a size of a few μm can be resolved and classified with our in-line holographic microscope HOLIMO. The algorithm for particle recognition proved to work reliably. Nevertheless, the number of particles that can be attributed correctly to their classes depends on several things. The most important one is choosing the right reconstruction plane separation in order to find the focus plane of the particle. This issue often leads to a frequency of occurrence with a lot of regular habits although there were no regular hydrometeors measured. Therefore, control of the outcome by eye becomes necessary. HOLIMO was improved with respect to the resolution. The results from the experiments described in this work revealed that it is more important to have a large numerical aperture instead of having a large geometrical magnification. Therefore, the most promising step towards increased resolution was to shorten the distance from the PS to the camera in order to decrease the error made by edge smearing. Here also a trade off was made because increasing the numerical aperture like this leads to a decrease of geometrical magnification and sharper objects could no longer be resolved when the fringe separation is too small for the camera to be recorded.

Hydrometeor habits have been investigated closely in different experiments at the AIDA about warm and mixed phase clouds and one focussing on a cirrus cloud. Data from the second AIDA campaign is in the evaluation phase and hence not represented in this thesis. The study of the three heterogeneous IN11 experiments concentrated on data sets about mixed phase clouds and cold clouds initiated from supercooled water droplets, ice seeds or sulfuric acid. The ice crystals were characterized by the frequency of occurrence of three habit categories (droplikes, regulars and irregulars). Those four experiments show the dependency of hydrometeor size and habit on temperature and the type of inorganic aerosol. The use of droplet and droplet plus sulfuric acid seeds in experiment 1 and 3 of the first AIDA campaign respectively lead to a similar characterization of the results. The frequency of occurrence of the 3 classes for hydrometeor sizes smaller than 60 μm identified the dominance of regular crystal shapes. WELAS showed for both experiments a high crystal concentration with a maximum of about 4700 cm^{-3}. The use of ice seeds additional to water seeds in experiment 2 of the same campaign presented an increased amount of irregular habits. Nevertheless, most of the ice crystal habits were still categorized as regular ice crystals. WELAS showed a low ice concentration with a maximum of about 200 cm^{-3}. Experiment 4 of the same campaign was conducted as a homogeneous freezing event for sulfuric acid. Under these circumstances one would expect hexagonal/regular ice crystal shapes. However, mainly aggregates were found, probably because of the high crystal concentration up to 100 cm^{-3} allowed ice crystals to aggregate.

The last campaign took place at the high altitude research station on the Jungfraujoch in Switzerland. The objective was to investigate the evolution of natural clouds and if there are patches of only droplets or only ice crystals. HOLIMO data was compared to a modelled well-mixed phase cloud with the same probability of ice crystals as found from HOLIMO. On one hand, the normalized frequency of ice crystals and cloud droplets vs. their abundance show the expected distribution for the modelled data set. Low numbers of ice crystals or cloud droplets occur with high frequencies, in fact it appears to be a 1/x relationship. On the other hand, the HOLIMO data set differs from this behavior for higher numbers of hydrometeors. This might indicate the presence of hydrometeor patches of one kind. Further investigations are in progress.

Acknowledgements

This work was supported by the ETH Zurich research grant TH-39/05-1 *'Development of an in-line Holographic Microscope for Ice Crystals'*. The authors are grateful to R. Schön from the Institute of Meteorology and Climate Research at the Forschungszentrum Karlsruhe, Germany for the help during the IN11 campaign in November 2007. The contributions to this thesis from M. Schnaiter, O. Möhler and S. Benz from the Institute for Meteorology and Climate Science at the Research Center Karlsruhe and from Evelyn Hesse from the University of Hertfordshire, Centre for Atmospheric and Instrumentation Research, Hatfield, Hertfordshire AL10 9AB, UK are gratefully acknowledged since they provided information and data from the IN11 and HALO02 campaigns in December 2008 during the whole period of this work. The author would like to express his gratitude to Felix Lüönd for his patience during and his time for discussing subject relevant matter. The author would also like to thank Felix Lüönd for his IDL routine to calculate ice supersaturations with respect to ice in % with the help of the Murphy-Koop parametrization. The author is furthermore very grateful to Edwin Hausamman, Hannes Wydler, Peter Isler and Hansjörg Frei. They always found the time when the author did not have any left in order to draw and build mechanical parts for the instrument used during this work. Peter Isler, Marc Wüest and Dani Lüthi are thanked for their help in infrastructural matters. The author would like to thank Stephane Laurent Gallavardin for lending and explaining his matlab routines in order to obtain fourier spectra from object reconstruction images via clustering. The authors gratefully acknowledge the NOAA Air Resources Laboratory (ARL) for the provision of the HYSPLIT transport and dispersion model and/or READY website (http://www.arl.noaa.gov/ready.html) used in this dissertation. The author is grateful to Dr. Daniel J. Cziczo who always had time for support on various occasions for instance on field campaigns. The authors would like to express their gratitude to Juranyi Zsofia and Ernest Weingartner from the Laboratory of Atmospheric Chemistry, Paul Scherrer Institut, Villigen, PSI, Switzerland for providing CPC data to the experiments carried out on the high altitude research station on the Jungfraujoch in Februar and March 2009.

Acknowledgements

I would like to thank Jörg Mäder for his kind and patient last minute help with statistic routines for R.

I would like to express my gratitude to Ulrike and Olaf who gave me the chance to work in a interesting field on a very interesting project.

I would like to thank Trude Storelvmo for her important feedback to one of the chapters.

I would like to thank the guys from the group for the fun time in the lab, office and during leisure time especially Felix, Marco (BM), Ingo, Cédric, Luis, Pia, Colombe, Gaby, Maria, Steffi, Sara, the two Anas, the two Hannas, Francesco, Vivek, Johanna and Valeria.

Last but not least I would like to thank my girlfriend Andrea for all her support over all these years. Her patience was my anchor in this turbid period of my life.

List of publications

Amsler, P. and Stetzer, O. and Schnaiter, M. and Hesse, E. and Benz, S. and Moehler, O. and Lohmann, U. (2009), Ice crystal habits from cloud chamber studies obtained by in-line holographic microscopy related to depolarization measurements, *Appl. Opt.*, accepted September 2009.

List of Tables

2.1 Linewidths and coherence lengths. 11

5.1 Geometrical magnification vs. resolution. 58

6.1 Frequency of occurrence of ice crystal habits in 3 different periods of experiment IN11_2 (see figures 6.3 and 6.4). 75

6.2 Frequency of occurrence of hydrometeor habits for three different size ranges for experiment 1. Size range I: 0-38 μm. Size range II: 38-76 μm. Size range III: 114-152 μm. The percentages are rounded figures. Therefore, it is possible that they add up to more or less than 100%. . . . 82

6.3 Frequency of occurrence of hydrometeor habits for three different size ranges for experiment 2. Size range I: 0-75 μm. Size range II: 75-150 μm. Size range III: 150-225 μm. The percentages are rounded figures. Therefore, it is possible that they add up to more or less than 100%. . . . 82

6.4 Frequency of occurrence of hydrometeor habits for one size range for experiment 3. Size range I from 0-60 μm. The percentages are rounded figures. Therefore, it is possible that they add up to more or less than 100%. 83

6.5 Frequency of occurrence of ice crystal habits for one size range for experiment 4. Size range I from 0-43 μm. The percentages are rounded figures. 85

6.6 Description of the 4 different IN11 experiments. 85

E.1 Basic reconstruction routine. 150

List of Figures

1.1   Ice crystal habit diagram for in cloud crystal growth (adapted from [200]).   4

1.2   Simplified shapes of possible ice crystals adapted from [202].   4

1.3   Measured critical supersaturation with respect to ice and temperature at which growth was observed for the different faces of a pristine ice crystal ([205]).   5

1.4   Multi layer nucleation growth related to an ice crystal habit growth diagram. Combined date from [208] and [204]. The condensation coefficient on the top panel shows a sharp local maximum of the basal face in the columnar regime between -4 an -10 °C on the lower panel and it indicates the transition between sheath and needle like ice crystal habits. This local maximum in the upper panel is followed by a local minimum towards increasing T. It indicates the transition from columnar to tabular ice crystal habits at -4°C on the lower panel. The condensation coefficient of the prism face on the top panel shows a broad maximum between -10 and -15°C indicating dendritical growth inside the tabular regime on the lower panel between -10 and -22°C for supersaturations with respect to ice around 20% and more.   6

2.2   Maximal path length difference (the d indicated in blue) for an in-line holographic microscope for coherent interference imaging between the object and sensor plane. A point on the object and sensor plane is given by the coordinates (x,y,z) and (X,Y,Z) respectively.   10

2.3 The figure shows in the image plane parts of a tilted text in the object space illuminated by the PS at position **B** situated at distance x_d+s from the hologram. An object at distance **s** is in focus at image distance v_s. **b** indicates the region illuminated by the PS on the screen. Region **c** is the circle in the image plane where all the points lying within the DOF in the object room will be displayed in focus, hence the name Circle of Confusion plane. It is calculated from the focal length divided by 1500 as a rule of thumb. 12

2.4 The influence of astigmatic aberration on the image obtained with an optical instrument is shown on this figure. It goes with the object position with respect to the central axis to the power of 2 in the power series expansion of the phase function. The entrance into the pupil **P** of the ray **W** coming from the object **O** will be pictured from the exit of the pupil **P'** into **O'**. In this case there will be two **O'** because of the inaptitude of representing points of lateral objects. Characteristic astigmatic aberrations of a spherical object are shown on the right hand side. 13

2.5 The Coma aberration and its influence on the picture quality is shown on this figure. The error goes with the object position with respect to the central axis to the power of 3 in the power series expansion of the phase function. This lateral asymmetry is shown on the right hand for a spherical object. The small point would be a coma free representation of the object. 14

2.6 Affection of the image quality by spherical aberration sketched on top of the figure. It goes with the object position with respect to the central axis to the power of 4 in the power series expansion of the phase function. It is an error of aperture because of its finite extension. The row of pictures at the bottom of the figure shows a positive spherical aberration of a PS. Images to the left are defocused toward the inside, images on the right toward the outside. 15

2.7 The possible distortions an optical instrument can show are sketched on this figure. They go with the object position with respect to the central axis to the first power in the power series expansion of the phase function. The left and right hand show pillow or barrel distortion of a lattice respectively. The lattice is displayed undistorted in the middle. 16

List of Figures

2.8 The influence of the wave field curvature on the focal plane is shown on this figure. The picture on the top shows the part of the reconstruction of several 5 μm PSL spheres farther away from the center of the reconstruction plane than those on the lower picture. The focal planes of those PSL spheres will be different because of this lateral discrepancy between the first and the second group on the image. 17

3.1 This figure depicts the different diffraction regimes in (a). The Fresnel number N_f in (b) indicates the region of interest. Φ_0 gives the divergence angle of the laser and W_0 its extension at the exit. 20

3.2 This figure explains the Huygens-Fresnel principle for reconstructions from holograms. The upper sketch shows the in-line PS setup for a digital holographic microscope. The reference wave interferes directly with the object at distance **l** from the PS producing the interference pattern at distance **L** from the PS on a digital camera. The lower panel indicates the plane inside V_{obs} where the object will scatter the reference wave. Elementary waves are excited by a plane wave front (blue) at the yellow points of the surface of an object. A convolution of all the scatterers result in the reconstructed wave front (green). 22

3.3 Sketch of two scatterers inside the observing volume of HOLIMO. 23

3.4 Wavefront behavior in the KH transformation. It takes place in the plane wavefront regime of the diverging wave. 23

3.5 This figure shows the paraxial approximation for in-line setups of optical instruments. The distance from a scatterer to the longitudinal z direction is a lot smaller than from the lateral y. The wave fields originating from the reference source, the object and the reconstruction source must have a small enough divergence. This means that the objects expansion in the y direction is a lot smaller than the distances p, r and s. If this holds then it holds in general also for the objects expansion in the x direction. The x axis is perpendicular to the y and z axes. 24

3.6 This figure shows the wavefront approximation made for the KH transformation. The distance of an object at the test point (x,y,z) from the center point (X,Y,Z) on the screen is given by $\vec{\xi} - \vec{r}$. In the approximation this becomes $\frac{\vec{\xi}\vec{r}}{\xi}$. 25

3.7 Scattered intensity distribution of a spherical aperture. The central spot caries 84% of the overall light intensity. 26

3.8 *Blooming and smearing of a hologram obtained with a CCD camera manifested in horizontal and vertical lines of overexposure.* 27

3.9 *Blooming and smearing of a picture obtained with a CCD camera for overexposure on the left and the same picture in normal exposure on the right operation.* 28

4.1 *Ability of focussing a laser beam in 3 dimensional space. D (laser source extension) is a lateral distance and F (distance between laser source and desired focal point) a longitudinal. A Gaussian beam can be collimated with D/F in lateral and $(D/F)^2$ in longitudinal distances.* 31

4.2 *Laser characteristics of a Gaussian beam. r indicates the point of focus of the laser beam. $2w_0$ is the point of minimal beam waist and θ gives the divergence angle of the beam.* 32

4.3 *Trigger considerations of HOLIMO. The camera and the laser need to be triggered at the same time. The laser pulse is initiated in the middle of the trigger signal. The camera needs to be exposed during a time period that covers the laser pulse.* 34

4.4 *This figure shows the Rayleigh resolution criterion for imaging systems. The top image shows the situation right in between the unresolved and the Rayleigh criterion case. They are sketched along the resolved case at the bottom of the figure.* 40

4.5 *The Influence of edge blurring on sharpness demonstrated on a droplet. The amount m of side lobes m of its interference pattern determines the slope $\frac{1}{w}$ of the transition from background to droplet ([210]).* 42

4.6 *The influence of edge blurring on a normalized rectangular function with respect to the amount of recorded interference fringes. All m reproduce the rectangle but the slope of the edges will increase with increasing m. Hence the size of the rectangle is more accurate with large m values ([210]).* 43

List of Figures

4.7 Examples for the thresholding selection process. The 53 spots represent one reconstruction plane within V_{obs} between PS and camera. A flat histogram (the difference and outline between the spots given in relative brightness is smaller than within others) like the one at the right hand side leads to a rejection of the parameters attributed to this reconstruction plane when there is a histogram like the one at the left hand side within the reconstruction sequence. 45

4.8 Images illustrating the working principle of auto focussing. The reconstruction sequence starts at the top left image and ends at the bottom right. Somewhere in between V_{obs} there is one plane of reconstruction with the maximum mean brightness. It is the focal plane of the object. . . . 46

4.9 Classification scheme by Magono et al. ([212]) 47

4.10 Ice crystal classification after Korolev ([213]). 48

4.11 Flow chart of the data processing of HOLIMO (left panel). First the hologram will be read in and a predefined routine reconstruction produces the image of maximal brightness at a distance l_j. The image needs to be binarized in order to define a boundary box. This makes it possible to classify the objects in a predefined routine and store the important findings. Every hologram is treated in the same manner before the data processing is ended. An example of this process is shown on the right hand side. Frame A shows the hologram, frame B its reconstruction and frame C its binary representation. Frame D includes the boundary box with the binary size label d_{equiv} underneath the box and frame E attributes the object to the class droplikes. 49

4.12 Classification scheme for HOLIMO hydrometeor images. The measurable parameters D_w, D_{max}, A and the circumference are used for image habit recognition. The aspect ratio $\alpha = D_w D_{max}^{-1}$ and the roundness $\beta = 4A(\pi D_{max}^2)^{-1}$ are given for 4 simple shapes. 50

List of Figures

4.13 Characterization of ice crystal habits with the help of a clustering algorithm. This figure shows 20 classes with different frequencies of occurrence. A lot of them are attributed to different backgrounds. This can be seen on the charts on the right hand side of the cluster classes. A size proxy from HOLIMO II shows the classes that are important for the characterization of the hydrometeors. The charts also show the time evolution of the hydrometeors inside each class and whether they were found in an ascending regime or not. 51

4.14 Figure drawing of HOLIMO I. 52

4.15 Sketch of the working principle of HOLIMO. It shows the recording setup. The reconstruction is done numerically. Basically the setup consists of a laser PS of light and a CCD camera that records the interference pattern of the reference wave with the scattered wave amplitude from an object inside the sample flow tube. A possible interference pattern on the camera sensor inside the light cone is shown on the right hand side of the sketch. Objects are sucked through the sample flow tube with the help of a vacuum pump. The mass flow controller (MFC) controls the flow. . . . 53

4.16 HOLIMO II at the AIDA facility in November 2007. Camera and laser are distributed to two boxes made out of thermal insulating material. The red arrows indicate the position of the camera, the laser, its power supply and the heaters. The measuring cell (not shown) would be attached to the camera on the other side of the insulating material. The laser is about 28 cm long. 55

4.17 Setup of HOLIMO II at the AIDA chamber in December 2008. The measuring cell, the camera, the laser, its power supply and the heaters are placed inside the aluminum box. There position is indicated with red arrows. 56

5.1 The UASF1951 target shows a descending helix of horizontal and vertical pairs of lines in groups and elements. The resolution limit for the highest pair of group and element number (7,6) is equal to about 4.4 μm. This highest group and element number pair can be identified on the reconstruction of the target to its right for a PS to CCD distance of 8 cm and a geometrical magnification of 9. 58

List of Figures

5.2 Different reconstruction of different 5 μm PSL spheres resulting in different resolutions of them. (a) shows the interference pattern of a group of 5 μm PSL spheres on the left, the distribution of the spheres is recognizable, and their reconstruction on the right. They show a thicker rim of the 5 μm PSL spheres than the reconstructed group of PSL spheres shown in (b). This causes the resolution to differ about 50% from 5 μm probably due to edge blurring. 60

5.3 Ice analogues seen with SEM and an optical microscope vs. holographic imaging. Panel (a) shows a SEM, a holographic reconstructed and an optical microscope image clockwise. The first row on panel (b) shows four reconstructions equivalent to the second row that shows images of an optical microscope from the same objects. 62

5.4 This figure shows reconstructed images from 20.2μm PSL spheres used for calibration. The spheres do not show a uniform distribution overall. Most of the observed PSL spheres were of the same size like the ones in the clusters on the right and left. Those 'regular' PSL (blue arrow) were sometimes accompanied by bigger ones (red arrow). The object indicated with the red arrow measures 27% more than the one indicated with the blue arrow. 63

5.5 Example pictures of a ZINC experiment classified with respect to temperature, ice supersaturation and experiment time. Interestingly, there seems to be an ice crystal with rosette like habits of about 30 μm. 65

6.1 Adiabatic expansion in the AIDA chamber. The sketch on top shows natural adiabatic process behavior. As long as the pressure is decreasing the temperature will decrease also. The two sketches at the bottom of this figure illustrate such processes and how they occur inside the AIDA chamber. The temperature levels off after a certain amount of time although the pressure is still decreasing. This is because of the influence of the walls. If the pressure is increasing then also the temperature increases. 68

6.2 Sketch of the AIDA facility. AIDA itself is the inner most cylinder. It has a diameter of 4 m and is 7 m high. This corresponds to a volume of approximately 84 m^3. It is surrounded by a thermal housing and aerosol and trace gas instruments. The wall temperature of AIDA is adjusted via heat exchange controlled by a cryostat in the basement. The temperature can be set between $-90°$ C and $+60°$ C. The inner temperature is controlled via adiabatic expansion with a vacuum system in the basement. The pressure can be set between 0.01 hPa and 1000 hPa. The vacuum system is also used for various sampling streams which are drawn off from the bottom of AIDA and controlled with a MFC. The point where HOLIMO was inserted into the measuring flow is indicated. 70

6.3 Combined results from AIDA mixed phase cloud experiment 2 of IN11. Panel a shows the temperature of the wall of AIDA and the gas it contains and also its pressure. Panel b shows the ice saturation ratio of the total and interstitial water content inside AIDA. Panel c (color bar in dN/dlogd$_p$) and d show the WELAS and HOLIMO size distribution respectively. Panel e shows the backward scattered signal of the perpendicular and the parallel channel in blue and red with respect to the total forward scattered signal in black. Panel f shows parts of the linear depolarization ratio of the 2 backward channels. The vertical lines throughout all panels indicate the time of droplet injection. 73

6.4 Ice crystal habits of experiment 2 of IN11 during three different time slots showing three different phases of habits and frequency of occurrences. . . 74

6.5 Example pictures of aspect ratios $\chi = maximumlength/thickness$ for thin and thick plates. The example on the left hand side of a thin plate seen under grazing incident has a χ of 24. The thin plate on the right hand side has an apparent χ of 15 though the contribution of edge blurring is somewhat bigger than for the previous example. This is due to the fact that the effect of both the forefront and its opposite add up. The thick plate has a comparable low χ of 2. 76

6.6 Linear depolarization ratios vs. χ. 77

6.7 Ray paths which contribute most to linear depolarization for $\chi < 1.5$ (a), $2 < \chi < 18$ (b) and $\chi > 18$ (c). 78

List of Figures

6.8 Time evolution of ice crystal habit, relative humidity, pressure and temperature profile of the MPC experiment 1 in AIDA at 255 K. The upper most plot shows the temperature (dashed line) and pressure evolution (gray line). The next lower plot shows the total number concentration $c_{n\ tot}$ in cm^{-3} (black points) recorded with WELAS for a size range from 5 μm to 200 μm. This instrument is an optical particle counter that is calibrated for spherical particles. This plot also shows the size proxy D_w in μm (gray points) from HOLIMO. It shows the time span of observable hydrometeors with WELAS and HOLIMO respectively. The smallest hydrometeor found with HOLIMO was 2 μm and the biggest 140μm. The gray shaded regions ($187 \leq t_{exp} \leq 806 s$ and $1682 \leq t_{exp} \leq 1757$ s) in the RH_{ice} profile indicate supercooled water droplet injection. Representative samples of HOLIMO images of different classes are shown underneath the RH profile. Predominantly droplike and regular hydrometeors were recorded with HOLIMO and they are shown independently with respect to time from the top left to the bottom right corner. 80

6.9 As figure 6.8, but for experiment 3 with sizes between 2 μm and 50 μm. Supercooled water droplet aerosols were injected during the experiment indicated by the green shaded region. Droplike, regular and irregular hydrometeors were recorded with hOLIMO and they are shown independently with respect to time from the left to the right corner. 81

6.10 As figure 6.8, but for experiment 4 with sizes between 2 μm and 43 μm. Only irregular hydrometeors were recorded with HOLIMO. They are shown with respect to time from the top left to the bottom right corner. . . 84

6.11 Frequency of occurrence of hydrometeor habits for sizes <60 μm for the 4 different IN11 experiments. 86

7.1 Illustrations from [226] about the patchiness of clouds. a is a fully mixed phase cloud and b shows regions where there exist only droplets and others where there exist only ice crystals. 90

7.2 This figure shows the HOLIMO (red arrow) instrument applied to the high altitude research station on the Jungfraujoch February and March 2009. It is placed next to the fog sampler in gold facing west. 91

7.3 Relative humidity, temperature, pressure and wind speed for March 1 2009 between 13:00 and 14:00 [courtesy of MeteoSchweiz]. 93

7.4 Back trajectory calculated with the on-line HYSPLIT model with respect to ambient temperature and relative humidity ([229]) for March 1 2009. . 94

7.5 Ice crystal habits from the PINC II campaign from March 1 2009 in the afternoon 13:33 - 13:46 local time. Pictures are shown in chronological sequence from top left to bottom right. 95

7.6 Modelled random distribution of consecutive ice crystals and cloud droplets (blue bars) vs. measured distribution (red bars) on the left and right panel respectively. The probability to measure ice crystals was measured to be 0.5 and therefore set to ≤ 0.5 in the modelled random distribution. 97

8.1 Comparison between binarization with respect to a certain threshold and smoothing without. A thin deformed triangle, a droplet and a thin hexagonal were investigated. The object on the right hand side should be a thin hexagonal plate. In this case both binarization via thresholding and smoothing failed to deliver a true image of the object. 99

8.2 3 D loop rotation around a reconstructed object obtained from a hologram around the inherent aperture angle of about $9°$. One can see that the plane shown is tilted towards the right in the upper left corner and is moved to the left towards the lower right corner. 100

8.3 3D modelling of a dendrite and a droplet. The dendrite is composed of all the parts in focus of the dendrite coming from different reconstruction planes in order to obtain a 3D model representation of the dendrite. The droplet is treated with isosurface and -caps in order to obtain a 3D model representation of the droplet. 102

C.1 This figure shows the operation of the source router on panel a). It switches between 2 channels. In the HOLIMO setup channel 1 is used. The physical channel fills automatically a surface of a cluster of surfaces described in b). This allows for simultaneous filling and analyzation. . . . 145

C.2 The state diagram of the trigger manager is responsible for generating the TE. The transition between the states are controlled by the events written in bold characters. The events generated during a transition are written in italic. 146

D.1 Program structure back panel of the snapshot_example2 vi for an ActiveX application. 148

List of Figures

E.1 *Negative vs. positive reconstruction images of a thin hexagonal plate.* . . 151

E.2 *Contrast image reconstructions of one hologram of a droplet out of and in focus. The reconstructions reveal some noise and a weak contrast overall.* 152

Bibliography

[1] O. Boucher. Air traffic may increase cirrus cloudiness. *Nature*, 397:30–31, 1999.

[2] C. S. Zerefos, K. Eleftheratos, D. S. Balis, P. Zanis, G. Tselioudis, and C. Meleti. Evidence of impact of aviation on cirrus cloud formation. *Atmos. Chem. Phys.*, 3:1633–1644, 2003.

[3] P. Minnis, J. K. Ayers, R. Palikonda, and D. Phan. Contrails, cirrus trends, and climate. *J. of Climate*, 17:1671–1685, 2004.

[4] A. J. Heymsfield and G. M. McFarquhar. High albedos of cirrus in the tropical pacific warm pool: Microphysical interpretations from CEPEX and from Kwajalein, Marshall islands. *J. Atmos. Sci.*, 53:2424–2445, 1996.

[5] K. Sassen and S. Benson. A mitlatitude cirrus cloud climatology from the facility of atmospheric remote sensing: II. Microphysical properties derived from lidar depolarization. *J. Atmos. Sci.*, 58:2103–2112, 2001.

[6] W. Haag, B. Kärcher, J. Ström, A. Minikin, U. Lohmann, J. Ovarlez, and A Stohl. Freezing thresholds and cirrus cloud formation mechanisms inferred from in situ measurements of relative humidity. *Atmos. Chem. Phys.*, 3:1791–1806, 2003.

[7] P. J. Demott, Y. Chen, S. M. Kreidenweiss, D. C. Rogers, and D. E. Sherman. Ice formation by black carbon particles. *Geophys. Res. Lett.*, 26:2429–2432, 1999.

[8] B. Zuberi, A. K. Bertram, C. A. Cassa, L. T. Molina, and M. J. Molina. Heterogeneous nucleation of ice in $(NH_4)_{(2)}SO_4$-H_2O particles with mineral dust immersions. *Geophys. Res. Lett.*, 29(1504):142–1:142–4, 2002.

[9] V. J. Schaefer. The production of ice crystals in a cloud of supercooled water droplets. *Science*, 104(2707):457–459, 1946.

[10] G. Vali. Ice nucleation-A review. *Nucleation and Atmospheric Aerosols*, page 18, 1996.

[11] U. Lohmann and J. Feichter. Global indirect aerosol effects: A review. *Atmos. Chem. Phys.*, 5:715–737, 2005.

[12] W. Cantrell and A. Heymsfield. Production of ice in tropospheric clouds - A review. *Bull. Am. Meteorol. Soc.*, 86(6):795–807, 2005.

[13] Y. Zhang, A. Macke, and F. Albers. Effect of crystal size spectrum and crystal shape on stratisform cirrus radiative forcing. *Atmos. Res.*, 52:59–75, 1999.

[14] J. E. Kristjánsson, J. M. Edwards, and D. L. Mitchell. Impact of a new scheme for optical properties of ice crystals on climate of two gcms. *J. Geophys. Res.*, 105:10063–10079, 2002.

[15] U. Lohmann, J. Zhang, and J. Pi. Sensitivity studies of the effect of increased aerosol concentrations and snow crystal shape on the snowfall rate in the arctic. *J. Geophys. Res*, 108:10063–10079, 2003.

[16] L. Levkov, M. Boin, and B. Rockel. Impact of primary ice nucleation parametrization on the formation and maintenance of cirrus. *Atmos. Res.*, 38:147–159, 1995.

[17] U. Lohmann and G. Lesins. Stronger constraints on the anthropogenic indirect aerosol effect. *Science*, 298(5595):1012–1015, 2002.

[18] S. A. Alexandrov, T. R. Hillman, T. Gutzler, and D. D. Sampson. Synthetic aperture fourier holographic optical microscopy. *Phys. Rev. Let.*, 97(16):4, 2006.

[19] W. L. Anderson and R. E. Beissner. Counting and classifying small objects by far-field light scattering. *Appl. Opt.*, 10(7):1503–&, 1971.

[20] D. A. Ansley and L. D. Siebert. Pulsed laser holography. *Ann. N.Y. Acad. Sci.*, 168(A3):475–&, 1970.

[21] G. Barbastathis and A. Sinha. Information content of volume holographic imaging. *Trends Biotechnol.*, 19(10):383–392, 2001.

[22] J. J. Barton. Removing multiple-scattering and twin images from holographic images. *Phys. Rev. Lett.*, 67(22):3106–3109, 1991.

[23] R. A. Belz and R. W. Menzel. Particle field holography at arnold-engineering-development-center. *Opt. Eng.*, 18(3):256–265, 1979.

[24] R. Bexon. Magnification in aerosol sizing by holography. *J. Phys. E: Sci. Instrum.*, 6(3):245–248, 1973.

[25] R. Bexon, G. D. Bishop, and J. Gibbs. Automatic assessment of aerosol holograms. *J. Aerosol Sci.*, 7:397–407, September 1976.

[26] R. Bexon and K. E. Cooke. Light emitting diode device for use in fault finding in pulsed laser holography. *J. Phys. E: Sci. Instrum.*, 7(7):511–512, 1974.

[27] R. Bexon, M. G. Dalzell, and M. C. Stainer. In-line holography and the assessment of aerosols. *Opt. Laser Technol.*, 8:161–165, August 1976.

[28] S. Borrmann and R. Jaenicke. Application of microholography for ground-based in-situ measurements in stratus cloud layers - a case-study. *J. Atmosp. Oceanic Technol.*, 10(3):277–293, August 1993.

[29] K. Bromley, M. A. Monahan, J. F. Bryant, and B. J. Thompson. Holographic subtraction. *Appl. Opt.*, 10(1):174–&, 1971.

[30] P. R. A. Brown. Use of holography for airborne cloud physics measurements. *J. Atmos. Oceanic Technol.*, 6(2):293–306, 1989.

[31] O. Bryngdahl and A. Lohmann. Single-sideband holography. *J. Opt. Soc. Am. A*, 58(5):620–&, 1968.

[32] S. L. Cartwright, P. Dunn, and B. J. Thompson. Noise and resolution in far-field holography. *J. Opt. Soc. Am. A*, 70(12):1631–1631, 1980.

[33] S. L. Cartwright, P. Dunn, and B. J. Thompson. Particle sizing using far-field holography. New developments. *Opt. Eng.*, 19(5):727–733, 1980.

[34] E. B. Champagne. Nonparaxial imaging magnification and aberration properties in holography. *J. Opt. Soc. Am. A*, 56(10):1448–&, 1966.

[35] E. B. Champagne. Optimization of optical systems. *Appl. Opt.*, 5(11):1843–&, 1966.

[36] E. B. Champagne. Transform relations in coherent systems. *Appl. Opt.*, 5(6):1088–&, 1966.

[37] E. B. Champagne. Nonparaxial imaging, magnification and aberration properties in holography. *J. Opt. Soc.f Am. A*, 57(1):51–55, 1967.

[38] E. B. Champagne. *A qualitative and quantitative study of holographic imaging*. PhD thesis, Ohio State University, 1967.

[39] E. B. Champagne and L. Kersch. Control of holographic interferometric fringe patterns. *J. Opt. Soc. Am. A*, 59(11):1535–&, 1969.

[40] E. B. Champagne and N. G. Massey. Resolution in holography. *Appl. Opt.*, 8(9):1879–1885, 1969.

[41] J. M. Cowley and D. J. Walker. Reconstruction from in-line holograms by digital processing. *Ultramicroscopy*, 6(1):71–76, 1981.

[42] J. S. Crane, S. L. Cartwright, and B. J. Thompson. Far-field holography of phase objects. *J. Opt. Soc. Am. A*, 72(12):1825–1825, 1982.

[43] E. Cuche, P. Marquet, and C. Depeursinge. Simultaneous amplitude-contrast and quantitative phase-contrast microscopy by numerical reconstruction of fresnel off-axis holograms. *Appl. Opt.*, 38(34):6994–7001, 1999.

[44] J. B. Develis. Theory and applications of holography. *Am. J. Phys.*, 36:370–370, April 1968.

[45] J. B. DeVelis, Jr. Parrent, G. B., and B. J. Thompson. Image reconstruction with fraunhofer holograms. *J. Opt. Soc. Am. A*, 56(4):423–427, 1966.

[46] P. Dunn and B. J. Thompson. Object shape and resolution in far-field holography. *J. Opt. Soc. Am. A*, 69(10):1402–1402, 1979.

[47] P. Dunn and B. J. Thompson. Object shape, fringe visibility, and resolution in far-field holography. *Opt. Eng.*, 21(2):327–332, 1982.

[48] P. Dunn and J. M. Walls. Absorption and phase in-line holograms - comparison. *Appl. Opt.*, 18(13):2171–2174, 1979.

[49] P. Dunn and J. M. Walls. Improved micro-images from in-line absorption holograms. *Appl. Opt.*, 18(3):263–264, 1979.

[50] H. M. A. El-Sum. *Reconstructed Wave-Front Microscopy*. PhD thesis, Stanford University, 1953.

[51] V. K. S. Feige and L. J. Balk. Calibration of a scanning probe microscope by the use of an interference-holographic position measurement system. *Meas. Sci. Technol.*, 14(7):1032–1039, 2003.

[52] M. E. Fourney, J. H. Matkin, and A. P. Waggoner. Aerosol size and velocity determination via holography. *Rev. Sci. Instrum.*, 40(2):205–&, 1969.

[53] Y. Frauel, E. Tajahuerce, M. A. Castro, and B. Javidi. Distortion-tolerant three-dimensional object recognition with digital holography. *Appl. Opt.*, 40(23):3887–3893, 2001.

[54] J. P. Fugal, R. A. Shaw, E. W. Saw, and A. V. Sergeyev. Airborne digital holographic system for cloud particle measurements. *Appl. Opt.*, 43(32):5987–5995, 2004.

[55] D. Gabor. A new microscopic principle. *Nature*, 161(4098):777–778, 1948.

[56] D. Gabor. Microscopy by reconstructed wave-fronts. *Proc. R. Soc. London, Ser. A*, 197(1051):454–487, 1949.

[57] D. Gabor. Character recognition by holography. *Nature*, 208(5009):422–&, 1965.

[58] D. Gabor. Holograms as optical elements. *J. Opt. Soc. Am. A*, 57(4):562–&, 1967.

[59] D. Gabor. Outlook for holography. *Optik*, 28(5):437–&, 1969.

[60] D. Gabor. Progress in holography. *Rep. Prog. Phys.*, 32(3):395–&, 1969.

[61] D. Gabor. Holography, 1948-1971. *Science*, 177:299–313, July 1972.

[62] D. Gabor and G. W. Stroke. Holography and its applications. *Endeavour*, 28(103):40–&, 1969.

[63] D. Gabor, G. W. Stroke, D. Brumm, Funkhous.A, and A. Labeyrie. Reconstruction of phase objects by holography. *Nature*, 208(5016):1159–&, 1965.

[64] D. Gabor, G. W. Stroke, R. Restrick, Funkhous.A, and D. Brumm. Optical image synthesis (complex amplitude addition and subtraction) by holographic fourier transformation. *Phys. Lett. A*, 18(2):116–&, 1965.

[65] J. Garcia-Sucerquia, R. Castaneda, and F. F. Medina. Fresnel-fraunhofer diffraction and spatial coherence. *Opt. Commun.*, 205(4-6):239–245, 2002.

[66] J. Garcia-Sucerquia, W. B. Xu, M. H. Jericho, and H. J. Kreuzer. Immersion digital in-line holographic microscopy. *Opt. Lett.*, 31(9):1211–1213, 2006.

[67] J. Garcia-Sucerquia, W. B. Xu, S. K. Jericho, P. Klages, M. H. Jericho, and H. J. Kreuzer. Digital in-line holographic microscopy. *Appl. Opt.*, 45(5):836–850, 2006.

[68] J. Garcia-Sucerquia, W. B. Xu, S. K. Jericho, P. Klages, M. H. Jericho, and H. J. Kreuzer. Digital in-line holographic microscopy. *Appl. Opt.*, 45(5):836–850, 2006.

[69] W. B. Gordon. Far-field approximations to kirchhoff-helmholtz representations of scattered fields. *IEEE Trans. Antennas Propag.*, AP23(4):590–592, 1975.

[70] W. Grabowski. Measurement of the size and position of aerosol droplets using holography. *Opt. Laser Technol.*, 15(4):199–205, 1983.

[71] J. A. Guerrero-Viramontes, D. Moreno-Hernandez, F. Mendoza-Santoyo, and M. Funes-Gallanzi. 3d particle positioning from ccd images using the generalized lorenz-mie and huygens-fresnel theories. *Meas. Sci. Technol.*, 17(8):2328–2334, 2006.

[72] P. Hariharan. Longitudinal distortion in images reconstructed by reflection holograms. *Opt. Commun.*, 17(1):52–54, 1976.

[73] P. Hariharan. Hologram recording geometry - its influence on image luminance. *Optica Acta*, 25(6):527–530, 1978.

[74] P. Hariharan and A. Selvarajan. Double focus systems for holography. *Opt. Commun.*, 4:392–394, February 1972.

[75] G. Haussmann and W. Lauterborn. Determination of size and position of fast moving gas-bubbles in liquids by digital 3-d image-processing of hologram reconstructions. *Appl. Opt.*, 19(20):3529–3535, 1980.

[76] K. Heinz, U. Starke, and J. Bernhardt. Surface holography with leed electrons. *Prog. Surf. Sci*, 64(3-8):163–178, 2000.

[77] J. C. Heurtley. Scalar rayleigh-sommerfeld and kirchhoff diffraction integrals - comparison of exact evaluations for axial points. *J. Opt. Soc. Am. A*, 63(8):1003–1008, 1973.

[78] R. Hickling. Scattering of light by spherical liquid droplets using computer-synthesized holograms. *J. Opt. Soc. Am. A*, 58(4):455–&, 1968.

[79] R. Hickling. Holography of liquid droplets. *J. Opt. Soc. Am. A*, 59(10):1334–1339, 1969.

[80] T. R. Hillman, S. A. Alexandrov, T. Gutzler, and D. D. Sampson. Microscopic particle discrimination using spatially-resolved fourier-holographic light scattering angular spectroscopy. *Opt. Express.*, 14(23):11088–11102, 2006.

[81] K. D. Hinsch and S. F. Herrmann. Holographic particle image velocimetry. *Meas. Sci. Technol.*, 15(4):3, 2004.

Bibliography

[82] H. E. Hwang and G. H. Yang. Far-field diffraction characteristics of a time-variant gaussian pulsed beam propagating from a circular aperture. *Opt. Eng.*, 41(11):2719–2727, 2002.

[83] G. Indebetouw, Y. Tada, J. Rosen, and G. Brooker. Scanning holographic microscopy with resolution exceeding the rayleigh limit of the objective by superposition of off-axis holograms. *Appl. Opt.*, 46(6):993–1000, 2007.

[84] B. Javidi and E. Tajahuerce. Three-dimensional object recognition by use of digital holography. *Opt. Lett.*, 25(9):610–612, 2000.

[85] B. Javidi, S. Yeom, I. Moon, and M. Daneshpanah. Real-time automated 3d sensing, detection, and recognition of dynamic biological micro-organic events. *Opt. Express.*, 14(9):3806–3829, 2006.

[86] S. K. Jericho, J. Garcia-Sucerquia, W. B. Xu, M. H. Jericho, and H. J. Kreuzer. Submersible digital in-line holographic microscope. *Rev. Sci. Instrum.*, 77(4):043706-1:043706–10, 2006.

[87] S. F. Johnston. Holography: From science to subcultures. *Opt. Phot. News*, 15(7):36–41, 2004.

[88] F. G. Kaspar. Power spectrum object flux transmittance by holographic techniques. *J. Opt. Soc. Am. A*, 59(3):359–&, 1969.

[89] F. G. Kaspar and R. L. Lamberts. Effects of some photographic characteristics on light flux in a holographic image. *J. Opt. Soc. Am. A*, 58(7):970–&, 1968.

[90] F. G. Kaspar, R. L. Lamberts, and C. D. Edgett. Comparison of experimental and theoretical holographic image radiance. *J. Opt. Soc. Am. A*, 58(9):1289–&, 1968.

[91] J. B. Keller. Geometrical theory of diffraction. *J. Opt. Soc. Am. A*, 52(2):116–&, 1962.

[92] T. Kimura. Coherent optical fiber transmission. *J. Lightwave Technol.*, 5:414–428, April 1987.

[93] P. Kirkpatrick and H. M. A. Elsum. Image formation by reconstructed wave fronts .1. physical principles and methods of refinement. *J. Opt. Soc. Am. A*, 46(10):825–831, 1956.

[94] W. D. Koek, N. Bhattacharya, J. J. M. Braat, T. A. Ooms, and J. Westerweel. Influence of virtual images on the signal-to-noise ratio in digital in-line particle holography. *Opt. Express*, 13(7):2578–2589, 2005.

[95] H. J. Kreuzer, H. W. Fink, H. Schmid, and S. Bonev. Halography of holes, with electrons and photons. *J. Microsc.*, 178:191–197, 1995.

[96] H. J. Kreuzer, M. J. Jericho, I. A. Meinertzhagen, and W. B. Xu. Digital in-line holography with photons and electrons. *J. Phys. A: Gen. Phys*, 13(47):10729–10741, 2001.

[97] H. J. Kreuzer, K. Nakamura, A. Wierzbicki, H. W. Fink, and H. Schmid. Theory of the point-source electron-microscope. *Ultramicroscopy*, 45(3-4):381–403, 1992.

[98] H. J. Kreuzer and R. A. Pawlitzek. Digital in-line holography. *Europhy. News*, 34(2):12, 2003.

[99] J. T. LaMacchia and J. E. Bjorkholm. Resolution of pulsed laser holograms. *Appl. Phys. Let.*, 12(2):45–&, 1968. 11.

[100] M. Lehmann. A simple holographic setup. *J. Opt. Soc. Am. A*, 56(10):1448–&, 1966.

[101] E. N. Leith. Recent results in holography. *Vacuum*, 16(6):314–&, 1966.

[102] E. N. Leith and Upatniek.J. Microscopy by wavefront reconstruction. *J. Opt. Soc. Am. A*, 55(5):569–&, 1965.

[103] H. Lichte. Electron holography approaching atomic resolution. *Ultramicroscopy*, 20(3):293–304, 1986.

[104] A. Lohmann. Optische einseitenbandübertragung angewandt auf das gabormikroskop. *J. Mod. Opt.*, 3:97–99, February 1956.

[105] M. Malek, D. Allano, S. Coetmellec, C. Ozkul, and D. Lebrun. Digital in-line holography for three-dimensional-two-components particle tracking velocimetry. *Meas. Sci. Technol.*, 15(4):699–705, 2004.

[106] E. Malkiel, O. Alquaddoomi, and J. Katz. Measurements of plankton distribution in the ocean using submersible holography. *Meas. Sci. Technol.*, 10(12):1142–1152, 1999.

[107] P. Marquet, B. Rappaz, P. J. Magistretti, E. Cuche, Y. Emery, T. Colomb, and C. Depeursinge. Digital holographic microscopy: a noninvasive contrast imaging technique allowing quantitative visualization of living cells with subwavelength axial accuracy. *Opt. Lett.*, 30(5):468–470, 2005.

[108] O. Matoba, T. J. Naughton, Y. Frauel, N. Bertaux, and B. Javidi. Real-time three-dimensional object reconstruction by use of a phase-encoded digital hologram. *Appl. Opt.*, 41(29):6187–6192, 2002.

[109] T. S. McKechnie. Reduction of speckle by a moving aperture - theory and measurement. *Optik*, 41(1):34–44, 1974.

[110] T. S. McKechnie. Reduction of speckle by a moving aperture - first-order statistics. *Opt. Commun.*, 13(1):35–39, 1975.

[111] T. S. McKechnie. Reduction of speckle in an image by a moving aperture - second-order statistics. *Opt. Commun.*, 13(1):29–34, 1975.

[112] M. Meier. Holographic imaging systems and their lens equivalents. *Antennas Propag. Soc. Int. Symp.*, 5:190, 1967.

[113] R. W. Meier. Magnification and third-order aberrations in holography. *J. Opt. Soc. Am. A*, 55(8):987–992, 1965.

[114] R. W. Meier. Cardinal points and novel imaging properties of a holographic system. *J. Opt. Soc. Am. A*, 56(2):219–&, 1966.

[115] R. W. Meier. Holographic-image types and their aberrations. *J. Opt. Soc. Am. A*, 56(10):1448, 1966.

[116] R. W. Meier. Optical properties of holographic images. *J. Opt. Soc. Am. A*, 57(7):895–&, 1967.

[117] H. Meng, G. Pan, Y. Pu, and S. H. Woodward. Holographic particle image velocimetry: from film to digital recording. *Meas. Sci. Technol.*, 15(4):673–685, 2004.

[118] V. Mico, Z. Zalevsky, P. Garcia-Martinez, and J. Garcia. Superresolved imaging in digital holography by superposition of tilted wavefronts. *Appl. Opt.*, 45(5):822–828, 2006.

[119] V. Mico, Z. Zalevsky, P. Garcia-Martinez, and J. Garcia. Synthetic aperture super-resolution with multiple off-axis holograms. *J. Opt. Soc. Am. A*, 23(12):3162–3170, 2006.

[120] G. Molesini, D. Bertani, and M. Cetica. In-line holography with interference filters as fourier processors. *Optica Acta*, 29(4):479–484, 1982.

[121] I. Moon and B. Javidi. Shape tolerant three-dimensional recognition of biological microorganisms using digital holography. *Opt. Express.*, 13(23):9612–9622, 2005.

[122] K. Murata, H. Fujiwara, and T. Asakura. Use of diffused illumination on in-line fraunhofer holography. *Jpn. J. Appl. Phys., Part 1*, 7(3):301–&, 1968.

[123] T. J. Naughton and B. Javidi. Compression of encrypted three-dimensional objects using digital holography. *Opt. Eng.*, 43(10):2233–2238, 2004.

[124] P. Naulleau, M. Brown, C. Chen, and E. Leith. Direct three-dimensional image transmission through single-mode fibers with monochromatic light. *Opt. Lett.*, 21(1):36–38, 1996.

[125] D. B. Neumann. Geometrical relationships between the original object and the two images of a hologram reconstruction. *J. Opt. Soc. Am. A*, 56(7):858–861, 1966.

[126] D. B. Neumann. Stabilization of holographic interference fringes by feedback control. *J. Opt. Soc. Am. A*, 56(10):1448–&, 1966.

[127] D. B. Neumann. Holography of moving scenes. *J. Opt. Soc. Am. A*, 57(11):1406–&, 1967.

[128] D. B. Neumann, C. F. Jacobson, and G. M. Brown. Holographic technique for determining phase of vibrating objects. *J. Opt. Soc. Am. A*, 59(4):474–&, 1969.

[129] R. B. Owen, D. Baumgardner, and R. Berger. Holographic calibration of cloud particle measurement systems. *J. Opt. Soc. Am. A*, 2(13):P88–P88, 1985.

[130] R. B. Owen, M. H. Johnston, and R. B. Lal. Holography in space - the spacelab-iii mission. *J. Opt. Soc. Am. A*, 1(12):1221–1221, 1984.

[131] G. B. Parrent and G. O. Reynolds. Resolution considerations in hologram process. *J. Opt. Soc. Am. A*, 55(11):1566–&, 1965.

[132] Jr. Parrent, G. B. and G. O. Reynolds. Space-bandwidth theorem for holograms. *J. Opt. Soc. Am.*, 56(10):1400–1401, 1966.

[133] Jr. Parrent, G. B. and B. J. Thompson. On the fraunhofer (far field) diffraction patterns of opaque and transparent objects with coherent background. *Optica Acta*, 11(3):183–194, 1964.

[134] K. W. Pavitt, M. C. Jackson, R. J. Adams, and J. T. Bartlett. Holography of fast-moving cloud droplets. *J. Phys. E: Sci. Instrum.*, 3:971–975, December 1970.

[135] T. C. Poon, T. Yatagai, and W. Juptner. Digital holography - coherent optics of the 21st century: introduction. *Appl. Opt.*, 45(5):821–821, 2006.

[136] S. M. F. Raupach, H. J. Vossing, J. Curtius, and S. Borrmann. Digital crossed-beam holography for in situ imaging of atmospheric ice particles. *J. Opt. A: Pure Appl. Opt.*, 8(9):796–806, 2006.

[137] G. O. Reynolds and J. B. Develis. Hologram coherence effects. *IEEE Trans. Antennas Propag.*, AP 15(1):41–48, 1967.

[138] R. B. Roberts, R. L. Lyon, M. Page, and R. P. Miskus. Laser holography - its application to study of behavior of insecticide particles. *J. Econ. Entomol.*, 64(2):533–&, 1971.

[139] L. Rosen. Focused-image holography with extended sources. *Appl. Phys. Let.*, 9:337–339, 1966.

[140] U. Schnars and W. Juptner. Direct recording of holograms by a ccd target and numerical reconstruction. *Appl. Opt.*, 33(2):179–181, 1994.

[141] U. Schnars and W. P. O. Juptner. Digital recording and numerical reconstruction of holograms. *Meas. Sci. Technol.*, 13(9):R85–R101, 2002.

[142] J. Sheng, E. Malkiel, and J. Katz. Single beam two-views holographic particle image velocimetry. *Appl. Opt.*, 42(2):235–250, 2003.

[143] J. Sheng, E. Malkiel, and J. Katz. Digital holographic microscope for measuring three-dimensional particle distributions and motions. *Appl. Opt.*, 45(16):3893–3901, 2006.

[144] F. M. Shofner, T. G. Russell, and R. Menzel. 2-dimensional spatial frequency analysis of fraunhofer hologram of a small opaque sphere. *Appl. Opt.*, 8(10):2043–&, 1969.

[145] L. D. Siebert. Front-lighted pulse laser holography. *Appl. Phys. Let.*, 11(10):326–&, 1967.

[146] G. W. Stroke. White-light reconstruction of holographic images using transmission holograms recorded with conventionally-focused images and in-line background. *Phys. Let.*, 23(5):325–&, 1966.

[147] G. W. Stroke, D. Brumm, Funkhous.A, A. Labeyrie, and R. C. Restrick. On absence of phase-recording or twin-image separation problems in gabor (in-line) holography. *Br. J. Appl. Phys*, 17(4):497–&, 1966.

[148] G. W. Stroke, D. Brumm, Funkhous.A, and R. Restrick. 3 advances in fourier-transform holography. *J. Opt. Soc. Am. A*, 55(11):1566–&, 1965.

[149] G. W. Stroke and D. G. Falconer. Attainment of high resolutions in wavefront-reconstruction imaging. *Phys. Let.*, 13(4):306–309, 1964.

[150] G. W. Stroke and D. G. Falconer. Attainment of high resolutions in holography by multi-directional illumination and moving scatterers. *Phys. Let.*, 15(3):238–&, 1965.

[151] G. W. Stroke, R. Restrick, Funkhous.A, and D. Brumm. Resolution-retrieving compensation of source effects by correlative reconstruction in high-resolution holography. *Phys. Let.*, 18(3):274–&, 1965.

[152] G. W. Stroke and R. C. Restrick. Holography with spatially noncoherent light - (x-ray diffraction microscopy - image synthesis - 3-dimensional photography - t/e). *Appl. Phys. Let.*, 7(9):229–&, 1965.

[153] E. Tajahuerce and B. Javidi. Encrypting three-dimensional information with digital holography. *Appl. Opt.*, 39(35):6595–6601, 2000.

[154] E. Tajahuerce, O. Matoba, and B. Javidi. Shift-invariant three-dimensional object recognition by means of digital holography. *Appl. Opt.*, 40(23):3877–3886, 2001.

[155] B. J. Thompson. Fraunhofer diffraction patterns of opaque objects with coherent background. *J. Opt. Soc. Am. A*, 53(11):1350–&, 1963.

[156] B. J. Thompson. Holography. *Soc. Photo-Optical Inst. Eng. J.*, 9(3):83–&, 1971.

[157] B. J. Thompson. Holographic particle sizing techniques. *J. Phys. E: Sci. Instrum.*, 7(10):781–788, 1974.

[158] B. J. Thompson. Applications of holography. *Rep. Prog. Phys.*, 41(5):633–674, 1978.

[159] B. J. Thompson and J. B. Develis. Introduction to coherent optics and holography. *Sae Trans.*, 77:68–&, 1968.

[160] B. J. Thompson, J. H. Ward, and W. R. Zinky. Application of hologram techniques for particle size analysis. *Appl. Opt.*, 6(3):519–&, 1967.

[161] B. J. Thompson and W. R. Zinky. Holographic detection of submicron particles. *Appl. Opt.*, 7(12):2426–&, 1968.

[162] J. D. Trolinger. Particle field holography. *Opt. Eng.*, 14(5):383–392, 1975.

[163] J. D. Trolinger. Airborne holography techniques for particle field analysis. *Ann. N.Y. Acad. Sci.*, 267(JAN30):448–459, 1976.

[164] J. D. Trolinger, R. A. Belz, and W. M. Farmer. Holographic techniques for study of dynamic particle fields. *Appl. Opt.*, 8(5):957–&, 1969.

[165] G. A. Tyler and B. J. Thompson. Fraunhofer holography applied to particle size analysis. a reassessment. *Optica Acta*, 23(9):685–700, 1976.

[166] Upatniek.J. Improvement of microimage quality in holography and other coherent optical systems. *J. Opt. Soc. Am.*, 56(10):1448–&, 1966.

[167] J. Upatnieks and C. D. Leonard. Holographic light line sight. *J. Opt. Soc. Am. A*, 69(10):1402–1402, 1979.

[168] J. C. Urbach and R. W. Meier. Properties and limitations of hologram recording materials. *Appl. Opt.*, 8(11):2269–&, 1969. 42.

[169] A. van den Bos and A. J. den Dekker. Ultimate resolution in the presence of coherence. *Ultramicroscopy*, 60(3):345–348, 1995.

[170] R. F. vanLigten and H. Osterberg. Holographic microscopy. *Nature*, 211(5046):283–&, 1966.

[171] C. S. Vikram. Accurate linewidth measurements in aperture limited in-line Fraunhofer holography. *J. Mod. Opt.*, 37(12):2047–2054, 1990.

[172] C. S. Vikram. Optimizing image quality in Fraunhofer holography with variable intensity reconstruction beam. *Optik*, 86(2):58–60, 1990.

[173] C. S. Vikram. Trends in far-field holography. *Opt. Lasers Eng.*, 13(1):27–38, 1990.

[174] C. S. Vikram. *Particle Field Holgraphy*, volume 11. Cambridge University Press, 18 edition, 1992.

[175] C. S. Vikram. Rayleigh versus Marechal spherical-aberration tolerance in in-line Fraunhofer holography. *Opt. Eng.*, 33(11):3715–3717, 1994.

[176] C. S. Vikram. Resolution limits due to primary aberrations in Fraunhofer holography. *Optik*, 99(1):29–31, 1995.

[177] C. S. Vikram. Image formation in detuned interference-filter-aided in-line Fraunhofer holography. *Appl. Opt.*, 35(32):6299–6303, 1996.

[178] C. S. Vikram. Image power for size analysis in in-line Fraunhofer holography. *Opt. Lett.*, 21(14):1073–1074, 1996.

[179] C. S. Vikram and M. L. Billet. Gaussian-beam effects in far-field in-line holography. *App. Opt.*, 22(18):2830–2835, 1983.

[180] C. S. Vikram and M. L. Billet. Magnification with divergent beams in Fraunhofer holography of object inside a chamber. *Optik*, 63(2):109–114, 1983.

[181] C. S. Vikram and M. L. Billet. In-line Fraunhofer holography at a few far fields. *Appl. Opt.*, 23(18):3091–3094, 1984.

[182] C. S. Vikram and M. L. Billet. Optimizing image-to-background irradiance ratio in far-field in-line holography. *Appl. Opt.*, 23(12):1995–1998, 1984.

[183] C. S. Vikram and M. L. Billet. Some salient features of in-line Fraunhofer holography with divergent beams. *Optik*, 78(2):80–83, 1988.

[184] C. S. Vikram and M. L. Billet. Aberration limited resolution in Fraunhofer holography with collimated beams. *Opt. Laser Technol.*, 21(3):185–187, 1989.

[185] C. S. Vikram and T. E. McDevitt. Simple determination of magnification due to recording configuration in particle field holography. *Appl. Opt.*, 28(2):208–209, 1989.

[186] C. S. Vikram and R. K. Sood. Time-average holography with thin phase recording materials. *Nouvelle Revue d'Optique Appliquee*, 3:85–88, March 1972.

[187] H. J. Vossing, S. Borrmann, and R. Jaenicke. In-line holography of cloud volumes applied to the measurement of raindrops and snowflakes. *Atmos. Res.*, 49(3):199–212, 1998.

[188] J. H. Ward and B. J. Thompson. In-line hologram system for bubble-chamber recording. *J. Opt. Soc. Am. A*, 57(2):275–&, 1967. 4.

[189] P. E. Werner. Accuracy in film-scanner intensity measurements. *Acta Crystallogr., Sect. A: Found. Crystallogr.-Crystal Physics Diffraction Theoretical and General Crystallography*, A 26:489–&, 1970.

[190] J. T. Winthrop and C. R. Worthington. Convolution formulation of fresnel diffraction. *J. Opt. Soc. Am. A*, 56(5):588–&, 1966.

[191] J. T. Winthrop and C. R. Worthington. Fresnel-transform representation of holograms and hologram classification. *J. Opt. Soc. Am. A*, 56(10):1362–&, 1966.

[192] W. K. Witherow. High-resolution holographic particle sizing system. *Opt. Eng.*, 18(3):249–255, 1979.

[193] W. Xu, M. H. Jericho, H. J. Kreuzer, and I. A. Meinertzhagen. Tracking particles in four dimensions with in-line holographic microscopy. *Opt. Lett.*, 28(3):164–166, 2003.

[194] W. Xu, M. H. Jericho, I. A. Meinertzhagen, and H. J. Kreuzer. Digital in-line holography of microspheres. *Appl. Opt.*, 41(25):5367–5375, 2002.

[195] S. Yeom, I. Moon, and B. Javidi. Real-time 3-d sensing, visualization and recognition of dynamic biological microorganisms. *Proc. IEEE*, 94(3):550–566, 2006.

[196] M. Zambuto and M. Lurie. Holographic measurement of general forms of motion. *Appl. Opt.*, 9(9):2066–&, 1970.

[197] F. Zernike. The concept of degree of coherence and its application to optical problems. *Physica*, 5:785–795, 1938.

[198] F. Zernike. The diffraction theory of aberrations. *J. Opt. Soc. Am. A*, 42(1):72–72, 1952.

[199] Y. Zhang, G. Pedrini, W. Osten, and H. J. Tiziani. Whole optical wave field reconstruction from double or multi in-line holograms by phase retrieval algorithm. *Opt. Express*, 11(24):3234–3241, 2003.

[200] K. G. Libbrecht. The physics of snow crystals. *Rep. Prog. Phys.*, 68(4):855–895, 2005.

[201] D. M. Murphy and T. Koop. Review of the vapour pressures of ice and supercooled water for atmospheric applications. *Q. J. R. Meteorol. Soc.*, 131:1539–1565, 2005.

[202] P. Yang, H. L. Wei, H. L. Huang, B. A. Baum, Y. X. Hu, G. W. Kattawar, M. I. Mishchenko, and Q. Fu. Scattering and absorption property database for non-spherical ice particles in the near- through far-infrared spectral region. *Appl. Opt.*, 44(26):5512–5523, 2005.

[203] D. Lamb and W. D. Scott. Linear growth-rates of ice crystals grown from vapor-phase. *J. Cryst. Growth*, 12(1):21–31, 1972.

[204] D. Lamb and W. D. Scott. Mechanism of ice crystal-growth and habit formation. *J. Atmos.c Sci.*, 31(2):570–580, 1974.

[205] J. Nelson and C. Knight. Snow crystal habit changes explained by layer nucleation. *J. Atmos. Sci.*, 55(8):1452–1465, 1998.

[206] K. G. Libbrecht. A critical look at ice crystal growth data. *ArXiv Condensed Matter e-prints*, page 21, November 2004.

[207] K. G. Libbrecht. Precision measurements of ice crystal growth rates. *ArXiv Condensed Matter e-prints*, page 22, August 2006.

[208] T. Kobayashi. The growth of snow crystals at low supersaturations. *Philos. Mag.*, 6(71):1363–1370, 1961.

[209] G. M. Hale and M. R. Querry. Optical-constants of water in 200-nm to 200-mum wavelength region. *Appl. Opt.*, 12(3):555–563, 1973.

[210] D. M. Robinson. A calculation of edge smear in far-field holography using a shortcut edge trace technique. *Appl. Opt.*, 9(2):496–497, 1970.

[211] H. Koehler. On Abbe's theory of image formation in the microscope. *Optica Acta*, 28(12):1691–1701, 1981.

[212] C. Magono and C. Lee. Meteorological classification of natural snow crystals. *J. Fac. Sci.*, 2:321–335, 1966. Hokkaido University Series VII.

[213] A. Korolev, G. A. Isaac, and J. Hallett. Ice particle habits in stratiform clouds. *Q. J. R. Meteorol. Soc.*, 126(569):2873–2902, 2000.

[214] Z. Ulanowski. Ice analog halos. *Appl. Opt.*, 44(27):5754–5758, 2005.

[215] O. Stetzer, B. Baschek, F. Lüönd, and U. Lohmann. The zurich ice nucleation chamber (ZINC)-a new instrument to investigate atmospheric ice formation. *Aerosol. Sci. Technol.*, 42:64–74, 2008.

[216] A. Welti, F. Lüönd, Stetzer O., and U. Lohmann. Influence of particle size on the ice nucleating ability of mineral dusts. *Atmos. Chem. Phys. Dicuss.*, 9:6929–6955, 2009.

[217] S. Benz, K. Megahed, O. Möhler, H. Saathoff, R. Wagner, and U. Schurath. T-dependent rate measurements of homogeneous ice nucleation in cloud droplets using a large atmospheric simulation chamber. *J. Photochem. Photobiol., A*, 176(1-3):208–217, 2005.

[218] M. Schnaiter, S. Benz, V. Ebert, T. Leisner, O. Möhler, R. W. Saunders, and R. Wagner. Influence of particle size and shape on the backscattering linear depolarization ratio of small ice crystals. *J. Atmos. Sci.*, 2009. to be submitted.

[219] H.C. van de Hulst. *Light scattering by small particles*. Dover Publications, Inc, New York, 1981.

[220] A. Macke, J. Mueller, and E. Raschke. Single scattering properties of atmospheric ice crystals. *J. Atmos. Sci.*, 53(19):2813–2825, 1996.

[221] K. Sassen. *Light scattering by nonspherical particles*. New York: Academic Press, 1999.

[222] J. Cozic, B. Verheggen, E. Weingartner, J. Crosier, K. N. Bower, M. Flynn, H. Coe, S. Henning, M. Steinbacher, S. Henne, M. C. Coen, A. Petzold, and U. Baltensperger. Chemical composition of free tropospheric aerosol for PM1 and coarse mode at the high alpine site Jungfraujoch. *Atmos. Chem. Phys.*, 8(2):407–423, 2008.

[223] U. Baltensperger, M. Schwikowski, D. T. Jost, S. Nyeki, H. W. Gaggeler, and O. Poulida. Scavenging of atmospheric constituents in mixed phase clouds at the high-alpine site Jungfraujoch part I: Basic concept and aerosol scavenging by clouds. *Atmos. Environ.*, 32(23):3975–3983, 1998.

[224] J. Cozic, B. Verheggen, S. Mertes, P. Connolly, K. Bower, A. Petzold, U. Baltensperger, and E. Weingartner. Scavenging of black carbon in mixed phase clouds at the high alpine site Jungfraujoch. *Atmos. Chem. Phys.*, 7(7):1797–1807, 2007.

[225] B. Verheggen, J. Cozic, E. Weingartner, K. Bower, S. Mertes, P. Connolly, M. Gallagher, M. Flynn, T. Choularton, and U. Baltensperger. Aerosol partitioning between the interstitial and the condensed phase in mixed-phase clouds. *J. Geophys. Res. [Atmos.]*, 112(D23), 2007.

[226] A. Korolev and G. A. Isaac. Relative humidity in liquid, mixed-phase, and ice clouds. *J. Atmos. Sci.*, 63(11):2865–2880, 2006.

[227] A. V. Korolev. Rates of phase transformations in mixed-phase clouds. *Q. J. R. Meteorol. Soc.*, 134(632):595–608, 2008.

[228] A. Korolev and P. R.. Field. The effect of dynamics on mixed-phase clouds: Theoretical considerations. *J. Atmos. Sci.*, 65(1):66–86, 2008.

[229] R.R. Draxler and G.D. Rolph. Hysplit (Hybrid single-particle lagrangian integrated trajectory) model access via NOAA ARL READY. *NOAA Air Resources Laboratory, Silver Spring, MD*, 2003. http://www.arl.noaa.gov/ready/hysplit4.html.

[230] G. D. Rolph. Real-time Environmental Applications and Display System (READY). *NOAA Air Resources Laboratory, Silver Spring, MD.*, 2003. http://www.arl.noaa.gov/ready/hysplit4.html.

[231] adapted from wikipedia

[232] adapted from lecture scripts

[233] some adapted parts from wikipedia

[234] adapted from google

[235] adapted from http://www.edmundoptics.com

[236] adapted from http://www.svs-vistek.de

[237] P. Amsler, O. Stetzer, M. Schnaiter, E. Hesse, S. Benz, O. Moehler and U. Lohmann. Ice crystal habits from cloud chamber studies obtained by in-line holographic microscopy related to depolarization measurements. *Appl. Opt.*, 48 (30), 5811–5822, 2009.

Appendix A

List of Acronyms

Appendix A. List of Acronyms

AIDA chamber	Aerosol Interactions and Dynamics in the Atmosphere chamber
AR coating	Anti Reflex coating
ARL	Air Resources Laboratory
ATD	Arizona Test Dust
CCD camera	Charge Coupled Device camera
CE	Couple Efficiency
CoC	Circle of Confusion
CPC	Condensation Particle Counter
DCT	Discrete Cosine Transformation
DOF	Depth Of Field
ECMWF	European Center for Medium-range Forecasts
HALO02	second HALO campaign at the Research Center Karlsruhe December 2008
HOLIMO	HOLographi Instrument for Microscopic Objects
HTRG	Hardware TRiGger
HYSPLIT	HYbrid Single-Particle Lagrangian Integrated Trajectory
IN11	Ice Nucleation campaign at the Research Center Karlsruhe November 200
ISO	International Organization for Standardization
KH transformation	Kirchhoff-Helmholtz transformation
LIDAR	Light Detection And Ranging
LED	Light Emitting Diode
MFC	Mass Flow Controler
MFD	Mode Field Diameter
NA	Numerical Aperture
ND filter	Neutral Density filter
NOAA	National Oceanic and Atmospheric Administration
PINC II	campaign at the High Altitude Research Station Jungfraujoch Februar 2009
PS	Point Source
PSI	Paul Scherrer Institute
PSL spheres	Poly Styrene Latex spheres
QE	Quantum Efficiency
READY	Real-time Environmental Applications and Display sYstem
RH	Relative Humidity
SEM	Scanning Electron Microscope
SIMONE	Scattering Intensity Measurements for the Optical Detection of Ice
SMPS	Scanning Mobility Particle Sizer
STRG	Software TRiGger
USAF target	United States Air Force target
vi	virtual instrument
WELAS	WihitE Light Aerosol Spectrometer
ZINC chamber	Zurich Ice Nucleation Counter chamber

Appendix B

List of Symbols

symbol	units	description
$2a$	[μm]	maximum particle diameter
c	[cm^{-3}]	particle concentration
d	[μm]	resolution limit
d_{equiv}	[μm]	sphere equivalent diameter
D_{max}	[μm]	maximum diameter parallel to streamlines of sample flow
D_w	[μm]	maximum diameter perpendicular to streamlines of sample flow
m		number of airy rings
m_0		wavelength dependent magnification
m_1		geometric magnification
M		overall magnification
N		number of far field distances
N_{pix}		amount of pixels
V_{obs}	[mm^3]	observing volume of HOLIMO
α		D_w/D_{max} if $D_w < D_{max}$ or D_{max}/D_w if $D_w > D_{max}$
δ		linear depolarization ratio
Δ_l	[μm]	reconstruction plane separation
Δ_x, Δ_y	[μm]	pixel size on the camera
Δ_X, Δ_Y	[nm]	pixel size on the reconstruction plane
λ	[nm]	wavelength of the light source
λ_{eff}	[nm]	wavelength with respect to the medium
χ		aspect ratio of length over thickness

Appendix C

Image acquisition

The image acquisition needs three elements. There is the camera, a frame grabber and the host PC memory. Figure C.1 illustrates the basic principle.

Memory buffers are called surfaces. The grabber automatically transfers the acquired image to the destination surface. The surface can be filled during analyzation of another surface. A set of surfaces is called a cluster (figure C.1).

The acquisition is controlled via a hardware and software trigger mode (HTRG and STRG) generating a trigger event (TE). The trigger manager introduces a delay be-

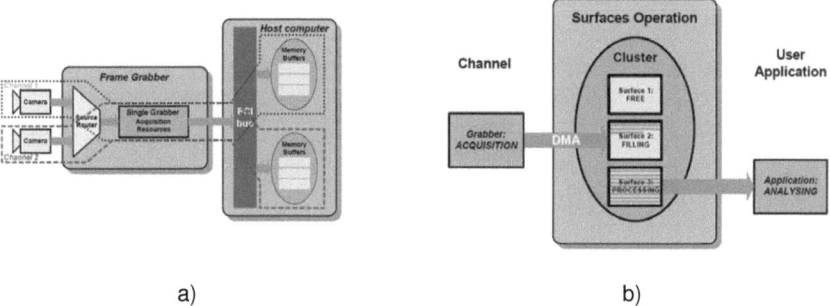

Figure C.1: *This figure shows the operation of the source router on panel a). It switches between 2 channels. In the HOLIMO setup channel 1 is used. The physical channel fills automatically a surface of a cluster of surfaces described in b). This allows for simultaneous filling and analyzation ([236]).*

Appendix C. Image acquisition

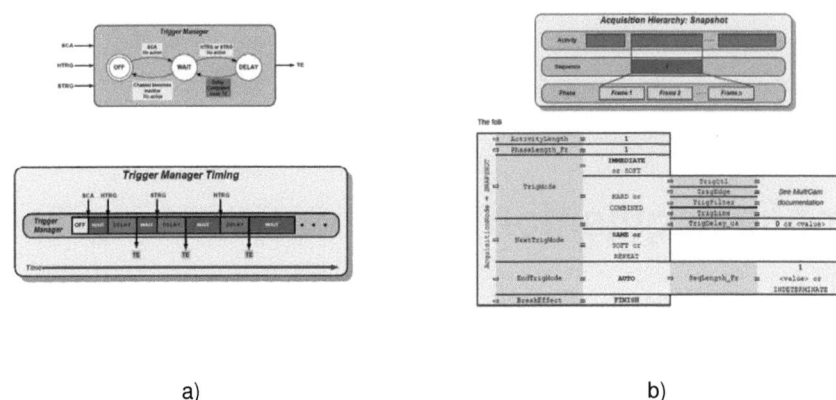

a) b)

Figure C.2: *The state diagram a) of the trigger manager is responsible for generating the TE which triggers the image aquisition b). The transition between the states are controlled by the events writte in bold characters. The events generated during a tran- characters. The events generated during a transition are written in italic ([236]).*

tween occurrence of a hardware trigger and a TE (figure C.1). The trigger manager has three states for this purpose. OFF means that the trigger manager is inactive. WAIT means that the trigger manager is waiting for a HTRG or STRG event and the DELAY state follows a HTRG or STRG event. The delay can be configured by the user.

With HOLIMO the immediate trigger and next trigger modes are used. This sets the delay to 0 and the sequence and slice managers do not wait for a TE (figure C.2). The acquisition mode is set to snapshot. It allows for controlled capturing of one picture with an area-scan camera.

Appendix D

Commercial software

The standard format for images with bit depths higher than 8 is the TIFF format and the standard way to store them is to produce an AVI stream. Along with the package of the camera and the laser came a simple software from SVS-Vistek. This AVI-recorder version 2007 ETH can provide such an AVI stream and allocate TIFF images from it. For this purpose one needs to scan the stream by eye. There is the possibility to upgrade this program such that it contains an inbuilt trigger for image recordings. Its working principle would make use of on-the-fly image analysis. Two different surfaces can be compared since surface operation allows for simultaneous surface filling and surface analysis. Therefore, a background image can be defined and a possible subsequent image can be compared with it. Possible ways to compare are mean intensities or a change of a certain amount of pixels. The software is based on the C-language but is copyright protected. The laser is controlled via the LabView interface UV-Q-Control-S. It allows for controlling the temperature of the crystal of the laser and hence also its output power to a certain extent. It also allows for running the laser at a certain frequency or triggered. There is also the small commercial C-routine ConvCam that helps programming the parameters of the camera such as exposure time and trigger mode for instance.

We tried to use the LabView platform in order to combine all those features. It was possible to program the camera in the ActiveX snapshot_example2.vi but we failed to make it fast enough for our application (D.1). Thus we stayed with Matlab.

148   Appendix D. Commercial software

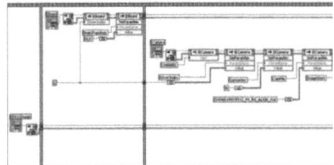
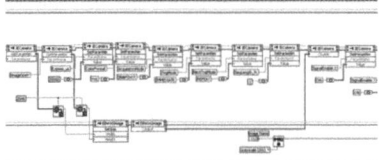
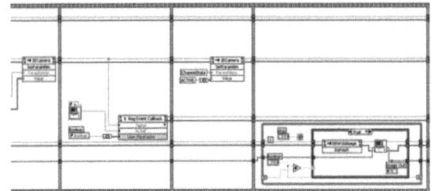
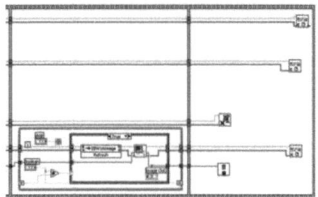

Figure D.1: *Program structure back panel of the snapshot_example2 vi for an ActiveX application.*

Appendix E

Basic reconstruction routine in Matlab

In this section the calculations for the reconstruction and characterization are described in greater detail. They were made in Matlab because this application offers a broad range of proceeding routines for images and straight forward treatments of them via matrix calculations. Before using a short basic reconstruction routine (code E.1) the data is read in and treated. This means that the user can trim different parameters to alter the results. According to figure 3.3, two bases are defined outside the basic reconstruction routine because they do not depend on the actual point of reconstruction. There are repeatable pattern coming from the kernel since the bases are symmetrical with respect to the center of the image. Therefore, as much as possible will be written in vector form in order to make use of the repmat function within Matlab and in order to avoid time consuming loops. Due to the weight in equation 3.2 there is a break of symmetry but there are still some data packages that can be duplicated. Along the reconstruction several variables are computed in order to identify and size the object. There were two routines used. The first and simple version allows for computing one or several reconstructions but it does not allow for classifying the results. In order to do so, several reconstructions are calculated and stored along with their average brightness. The maximum and minimum of all those values are giving an upper and a lower limit for the threshold of the segmentation. The focal distance l comes along with the maximum value and the reconstructed image of the object at this very point. The numerical magnification or the digital zoom factor can be changed inside the basic reconstruction routine by changing the pixel pitch of the reconstruction plane.

Basis1 is (-n+1:2:n-1)/2 which is (-3 -1 1 3) in the case of a 4x4 matrix where n accords with the resolution of the hologram. norm1 contains all the entries of basis1 squared. norm2 contains norm1 and $L^2 \cdot pitch^{-2}$ which is the square of the parameter L scaled by the area of the pixel in the original hologram plane plane. Weight is equal

Table E.1: Basic reconstruction routine.

```
for s = 1:n
    norm3 = repmat(1./sqrt(norm1+norm2(s)), n, 1);
    mat1 = repmat(im(s,:),n,1).*exp(norm3.*weight);
    mat2 = exp(((f./sqrt(norm1+norm2(s))*basis2(s))') * basis2);
    contraction = contraction + mat1*mat2;
end
```

to $(\delta_x * basis1' * basis1 + nom + l1\_delta\_vect) * conj(f)$ where δ_x is the pixel pitch in the reconstruction plane, nom is $L \cdot l^{-1}$ scaled by the pixel pitch in the original hologram plane and l1_delta_vect allows for tilting the reconstruction plane with respect to the hologram plane. Finally, the contraction adds up all the n^2 terms that are needed for the value of one reconstructed pixel.

E.1 Negative vs. positive reconstructions

It is a matter of taste whether the characterization of the object will be done from a negative or a positive reconstruction. Figure E.1 shows no difference between the negative and the positive reconstruction of a thin hexagonal plate. Maybe the rim of the hexagonal plate is better visible in the positive reconstruction. In this work, positive reconstructions are considered only. A second routine was programmed in order to classify the objects on the images with respect to their size and roundness (figure 6.1) in two different ways. The first way calculates the reconstruction again with respect to l without a background image in order to make a plain reconstruction from a hologram.

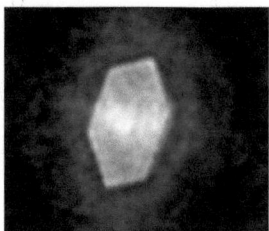
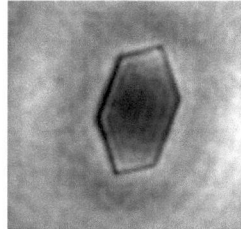

Figure E.1: *Negative vs. positive reconstruction images of a thin hexagonal plate.*

E.2 Contrast image

The second way calculates the reconstruction from the already obtained images with the background taken into account according to equation E.1.

$$I - |\vec{A}_{ref}|^2. \tag{E.1}$$

Figure E.2: *Contrast image reconstructions of one hologram of a droplet out of and in focus. The reconstructions reveal some noise and a weak contrast overall.*

Sometimes it was useful to start the calculation with the contrast image to obtain a bigger dynamic range and sometimes this lead to a very noisy picture due to a big change in background itself. Figure E.2 shows contrast image reconstructions from a droplet at different positions inside the observing volume. The background calculation takes care of the Gaussian intensity distribution by flattening it but exhibit noisy parts in each reconstructed image. Out of focus planes exhibit noise as well as in focus planes. This means that binarization via thresholding is still necessary but more difficult

E.2. Contrast image

because noise, background and object values are closer together. If the contrast in the interference pattern is weak than the contrast in the reconstructed image with removed background will be weak as well. Also, the background removal has to be done by hand because for every hologram containing an objects interference pattern a close enough background hologram needs to be taken into account of the calculation.

The characterization in this second routine was partitioned into preexisting supersaturations versus temperature, versus temperature and size and versus temperature, size and class. Additionally, size within a class and preexisting supersaturations versus temperature, regime and class were determined. The regime indicates whether the hydrometeor was found within an ascending or descending branch of the supersaturation.

As the routine methods worked very well for the first campaign in December 2007 at AIDA in Karlsruhe no further improvements have been made for the subsequent campaigns. This is due to the possible change of several parameters within the reconstruction routine. If, for instance, the aperture is widened by decreasing the distance between PS and camera (the only possible way to increase the aperture in this setup) then the required resolution for the reconstruction needs to be bigger. Another possible improvement would be to make the pixel pitch of the reconstruction plane smaller. This would cause the routine to zoom in on the picture. This also means that the real size of the reconstruction plane would be smaller and hence also the fringe visibility. Additionally, all objects outside this zoomed section would be omitted. Above all, the intensity distributions are very different for different aperture sizes.

Die VDM Verlagsservicegesellschaft sucht für wissenschaftliche Verlage abgeschlossene und herausragende

Dissertationen, Habilitationen, Diplomarbeiten, Master Theses, Magisterarbeiten usw.

für die kostenlose Publikation als Fachbuch.

Sie verfügen über eine Arbeit, die hohen inhaltlichen und formalen Ansprüchen genügt, und haben Interesse an einer honorarvergüteten Publikation?

Dann senden Sie bitte erste Informationen über sich und Ihre Arbeit per Email an *info@vdm-vsg.de*.

Sie erhalten kurzfristig unser Feedback!

VDM Verlagsservicegesellschaft mbH
Dudweiler Landstr. 99
D - 66123 Saarbrücken
www.vdm-vsg.de

Telefon +49 681 3720 174
Fax +49 681 3720 1749

Die VDM Verlagsservicegesellschaft mbH vertritt

Printed by Books on Demand GmbH, Norderstedt / Germany